# Contents

# A Handbook on the Requirements for Electrical Installations

## IEE Wiring Regulations
## 16th Edition

Based on BS 7671: 1992 incorporating
Amendment 1: 1994 and the 1995 Corrigenda

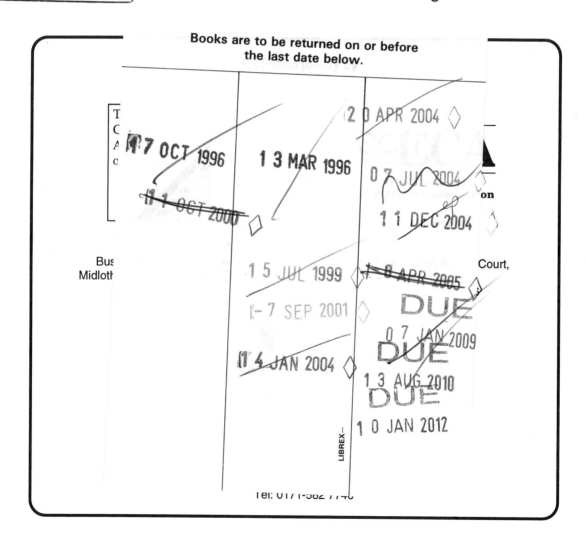

**Blackwell Science**

Copyright © The Electrical Contractors' Association,
The Electrical Contractors' Association of Scotland,
The National Inspection Council for Electrical
Installation Contracting 1991, 1995

Blackwell Science Ltd
Editorial Offices:
Osney Mead, Oxford OX2 0EL
25 John Street, London WC1N 2BL
23 Ainslie Place, Edinburgh EH3 6AJ
238 Main Street, Cambridge
    Massachusetts 02142, USA
54 University Street, Carlton
    Victoria 3053, Australia

Other Editorial Offices:
Arnette Blackwell SA
    1, rue de Lille, 75007 Paris
    France

Blackwell Wissenschafts-Verlag GmbH
    Kurfürstendamm 57
    10707 Berlin, Germany

    Feldgasse 13, A-1238 Wien
    Austria

First published by Blackwell
    Scientific Publications under the
    title *A Handbook on the 16th
    Edition of the IEE Regulations for
    Electrical Installations* 1991
Reprinted 1991 (three times), 1992
    (twice)
First edition revised published by
    Blackwell Science 1995

Printed and bound in Great Britain by
Hartnolls, Bodmin, Cornwall

DISTRIBUTORS

Marston Book Services Ltd
PO Box 87
Oxford OX2 0DT
(*Orders*: Tel: 01865 791175
        Fax: 01865 791927
        Telex: 837515)

USA
    Blackwell Science, Inc.
    238 Main Street
    Cambridge, MA 02142
    (*Orders*: Tel: 800 215-1000
            617 876-7000
            Fax: 617 492-5263)

Canada
    Oxford University Press
    70 Wynford Drive
    Don Mills
    Ontario M3C 1J9
    (*Orders*: Tel: 416 441-2941)

Australia
    Blackwell Science Pty Ltd
    54 University Street
    Carlton, Victoria 3053
    (*Orders*: Tel: 03 347-5552)

A catalogue record for this title
is available from the British Library

ISBN 0-632-03972-8 (pbk)
ISBN 0-632-04035-1 (looseleaf)

Library of Congress
Cataloging-in-Publication Data

A Handbook on the requirements for
    electrical installations (first edition revised).
        ISBN 0-632-03972-8 (pbk.). –
        ISBN 0-632-04035-1 (looseleaf version)
        1. Electric wiring – Handbooks, manuals, etc. 2.
    Electric wiring – Insurance requirements. I. Electrical
    Contractors' Association of Scotland. II. Electrical
    Contractors' Association (Great Britain). III. National
    Inspection Council for Electrical Installation Contracting
    (Scotland). IV. Institution of Electrical Engineers.
    Regulations for electrical installations.
    621.319′24′021841 – dc20            91-11754
                                            CIP

This publication was created using Ventura Publisher™

This edition technical authorship by John Peacock,
Building Systems Associates, 55 Pasture Road,
Wembley, Middlesex, HA0 3JW. Tel: 0181-904-5358

This edition page layout by Archetype, Stow-on-the-Wold

# Using the Handbook

This Handbook is arranged in six parts, including an index.

**PART A**

Introduction and Plan of BS 7671: 1992.

**PART B**

Index charts showing the arrangement of each chapter, accompanied where necessary by Points of Special Note. These charts also serve as an outline index to enable the reader to quickly identify Regulations relevant to his needs.

**PART C**

Topic charts, in which particular topics, e.g. isolation and switching, are presented as a co-ordinated reference to all the Regulations relevant to the subject, irrespective of the chapter in which they appear. Each topic chart is followed by supporting text, and is in one of two forms:

    (i)   decision charts, where the reader is led to the decision appropriate to his requirements; and

    (ii)   information charts and diagrams, where the reader is shown the access to the information for selecting the method or equipment most suitable.

**PART D**

Other illustrations for particular applications, indicating the requirements in visual form, together with explanatory notes.

**PART E**

Other useful sources of further information.

**PART F**

Index

NOTES

1. Parts B, C and D are cross-referenced as necessary.

2. It is anticipated that Parts A, B, C and E will be of greatest interest to the designer, whereas Parts C and D may be of more interest to the installer. The whole Handbook will be of use to college lecturers and students who require to study the Regulations in depth.

3. Regulation numbers throughout the book refer to BS 7671: 1992.

The co-operating bodies gratefully acknowledge the permission of the Institution of Electrical Engineers to quote from BS 7671: 1992.

## Summary of Illustrations and Diagrams

# A

## Introduction and Plan of BS 7671: 1992

# Introduction and Plan of BS 7671:1992 Including Amendment No 1 December 1994

## General

After fourteen editions of the Regulations for the Electrical Equipment of Buildings had been produced by the Institution of Electrical Engineers over a period of a hundred years, the 15th Edition – retitled Regulations for Electrical Installations, was introduced in 1981. After several sets of Amendments to the Regulations, various alterations to British Standards and Codes of Practice, and changes in legislation, a 16th Edition was published in May 1991. In October of 1992 the 16th Edition became a British Standard 7671:1992.

This edition of the Handbook is a revised version updated to reflect the changes to BS 7671:1992 arising from amendment No 1.

Many references are made in the Regulations to British Standards and Codes of Practice. Several aspects of electrical installation work, however, are specifically excluded from the Regulations, other than for their interface at power supplies, e.g. Emergency Lighting and Fire Alarms. Lightning protection and work in hazardous atmospheres are also excluded. In these areas, therefore, the relevant Standards and Codes provide authoritative information and also refer to the relevant Wiring Regulations, they should be readily available to designers and specifiers. Other aspects, e.g. work in mines and quarries, are the subject of separate legislation, as are explosives factories. Health and Safety legislation enters into almost every activity: designers and others should be aware of the wide range of informative publications produced by HSE.

The Regulations are divided into distinct parts: a statement of the hazards inherent in the use of electricity (danger from electric shock, fire, burns and injury from mechanical movement), a description of the various methods of combating these, the application of these methods, and Appendices some of which contain guidance on standard arrangements which will satisfy the Regulations in certain circumstances. The current-carrying capacity tables are also contained in the Appendices.

It will be seen that BS 7671:1992 cannot be used, or even read easily, in serial form. It is necessary to follow a topic through various parts, chapters and sections, in order to achieve understanding.

Recognizing that Designers, Installers, Inspectors and Testers need a minimum accepted standard there are Guidance Notes published by the IEE which should be obtained and used in conjunction with a study of BS 7671:1992.

This Handbook is intended to guide the designer and installer through BS 7671:1992 showing how each part is related to the whole, and to give practical guidance on how to approach new installations, extensions to existing installations, and the more extensive testing and inspection which is required.

## Amendment Number 1

This was published in December 1994 to become effective on 1st July 1995. In the main the amendments have been made to take account of EU CENELEC changes including the changes to the nominal supply voltage from 240 volts to 230 volts effective from 1st July 1995. Further amendments were made to align with CENELEC harmonisation document 384 and to correct grammar and punctuation errors in the previous issue.

Significant changes come into the sections relating to isolation and switching, protective extra-low voltage (PELV), swimming pools, electircal installations in caravans and for caravan parks (reversion to the term parks in place of sites as in the original issue). Extensive revision and updating of the appendices has taken place particularly as they relate to external influences and the forms of completion and inspection.

Within this handbook only amendments affecting regulations have been identified. Where there is a correction for punctuation or grammar, or a change to a BS number these have just been included in the text. Identification of changes in the text is by side bars and in diagrams or flow charts by underlining.

**Arrangement of BS 7671:1992**

BS 7671:1992 consists of Seven Parts, within each of which there are Chapters on the general subject, and Sections on particular aspects. Since the main objective has been to bring international accord to the work of electrical installation, future amendments will cover progress in the field. The division into Parts, Chapters and Sections is intended to facilitate the inclusion of such amendments and extensions when agreed.

The numbering system used for individual regulations in BS 7671: 1992 is significant, and assists in cross referencing.

| | |
|---|---|
| First digit | signifies the PART in which the Regulation is to be found |
| Second digit | signifies the CHAPTER of the Part |
| Third digit | signifies the SECTION of that Chapter |

Subsequent digits identify the particular Regulation number in that section. For example, Regulation 522 - 03 - 01 is to be found in:

| | |
|---|---|
| Part 5 | which deals with Selection and Erection of Equipment |
| Chapter 2 | which deals with Selection and Erection of Wiring Systems |
| Section 2 | dealing with external influences |
| Sub-section 03 | which deals with the presence of water (AD) or high humidity (AB) |

This relationship is shown diagrammatically in Figure A1, and it is suggested that oral reference to a Regulation should be made thus:

FIVE     TWO     TWO     DASH     ZERO     THREE     DASH     ZERO     ONE

This enables the listener to clearly identify the Part, Chapter, Section and Regulation.

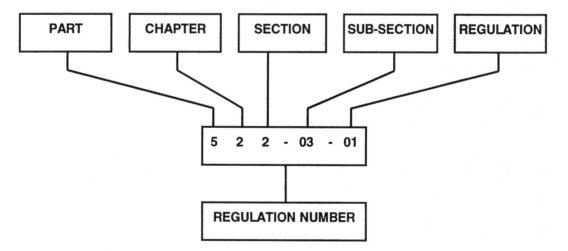

**Figure A1: Derivation of Regulation Numbers**

The plan of BS 7671:1992 is shown in two different ways in Figures A2 and A3:

   (i)  Figure A2 shows the plan by subject headings.

   (ii)  Figure A3 shows the interdependence of the various parts.

It will be seen that in Figure A3 two parts of the Regulations are shown connected by flow lines to other parts. These are Part 1 on the Scope, Object and Fundamental Requirements for Safety, and Part 2 giving Definitions. An understanding of these parts is essential to the use of the Regulations, and they are dealt with below.

The remaining parts are so interdependent that they are dealt with under their respective headings to discuss the principles, and by means of charts to identify topics which require reference to more than one part.

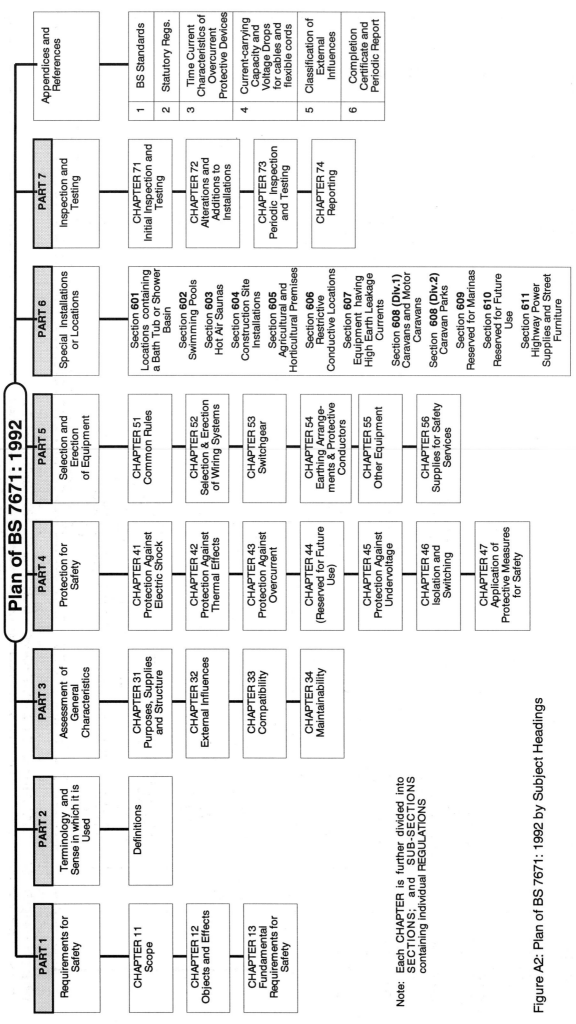

# Plan of BS 7671: 1992

**PART 1** — Requirements for Safety

- CHAPTER 11 Scope
- CHAPTER 12 Objects and Effects
- CHAPTER 13 Fundamental Requirements for Safety

**PART 2** — Terminology and Sense in which it is Used

- Definitions

**PART 3** — Assessment of General Characteristics

- CHAPTER 31 Purposes, Supplies and Structure
- CHAPTER 32 External Influences
- CHAPTER 33 Compatibility
- CHAPTER 34 Maintainability

**PART 4** — Protection for Safety

- CHAPTER 41 Protection Against Electric Shock
- CHAPTER 42 Protection Against Thermal Effects
- CHAPTER 43 Protection Against Overcurrent
- CHAPTER 44 (Reserved for Future Use)
- CHAPTER 45 Protection Against Undervoltage
- CHAPTER 46 Isolation and Switching
- CHAPTER 47 Application of Protective Measures for Safety

**PART 5** — Selection and Erection of Equipment

- CHAPTER 51 Common Rules
- CHAPTER 52 Selection & Erection of Wiring Systems
- CHAPTER 53 Switchgear
- CHAPTER 54 Earthing Arrangements & Protective Conductors
- CHAPTER 55 Other Equipment
- CHAPTER 56 Supplies for Safety Services

**PART 6** — Special Installations or Locations

- Section 601 Locations containing a Bath Tub or Shower Basin
- Section 602 Swimming Pools
- Section 603 Hot Air Saunas
- Section 604 Construction Site Installations
- Section 605 Agricultural and Horticultural Premises
- Section 606 Restrictive Conductive Locations
- Section 607 Equipment having High Earth Leakage Currents
- Section 608 (Div.1) Caravans and Motor Caravans
- Section 608 (Div.2) Caravan Parks
- Section 609 Reserved for Marinas
- Section 610 Reserved for Future Use
- Section 611 Highway Power Supplies and Street Furniture

**PART 7** — Inspection and Testing

- CHAPTER 71 Initial Inspection and Testing
- CHAPTER 72 Alterations and Additions to Installations
- CHAPTER 73 Periodic Inspection and Testing
- CHAPTER 74 Reporting

**Appendices and References**

| | |
|---|---|
| 1 | BS Standards |
| 2 | Statutory Regs. |
| 3 | Time Current Characteristics of Overcurrent Protective Devices |
| 4 | Current-carrying Capacity and Voltage Drops for cables and flexible cords |
| 5 | Classification of External Influences |
| 6 | Completion Certificate and Periodic Report |

Note: Each CHAPTER is further divided into SECTIONS; and SUB-SECTIONS containing individual REGULATIONS

Figure A2: Plan of BS 7671: 1992 by Subject Headings

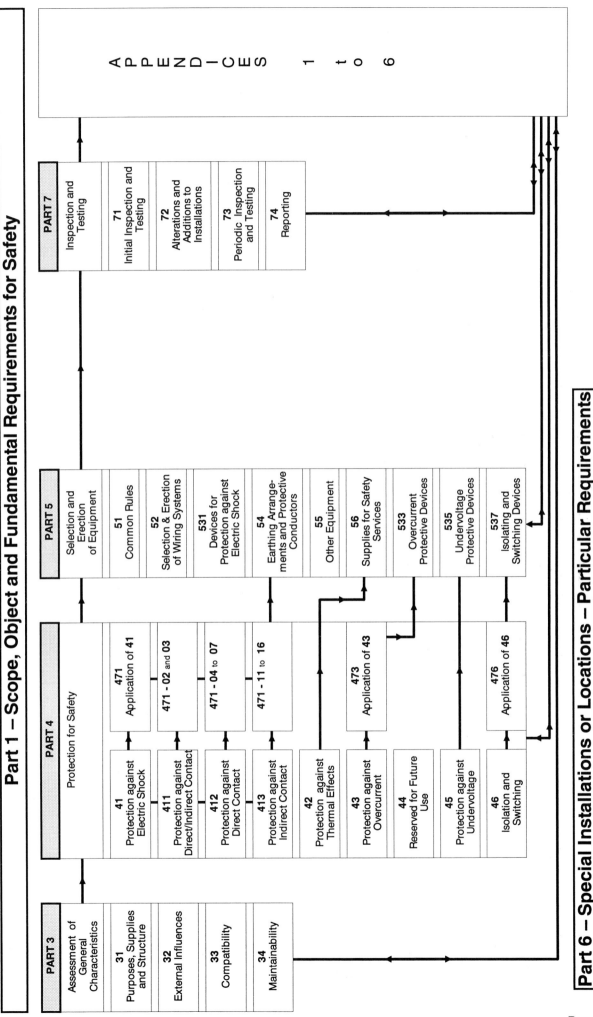

Figure A3: The Plan of BS 7671: 1992 showing Relationship of the Parts of the Regulations

## Guide to Interpreting Flow Charts

◯    Input point        ◯    Reference point

▢    Instruction point       ◇    Decision point

⬇    Continuation point      ⬆    Continuation point

## Symbols used Throughout the Handbook

| | | |
|---|---|---|
| $A^2s$ | = | Amperes-squared per second |
| $C$ | = | correction factor general |
| $C_a$ | = | correction factor for ambient temperature |
| $C_d$ | = | correction factor for type of overcurrent protective device |
| | | $C_d = $ 1 for HBC fuses and mcbs |
| | | $C_d = $ 0.725 for semi-enclosed fuses |
| $C_g$ | = | correction factor for grouping |
| $C_i$ | = | correction factor for conductors embedded in thermal insulation |
| $C_t$ | = | correction factor for operating temperature of conductor |
| $I$ | = | current (general term) |
| $I_a$ | = | current causing automatic operation of protective device within the time stated |
| $I_b$ | = | design current of a circuit |
| $I_d$ | = | fault current of first fault (IT system) |
| $I_n$ | = | nominal current or current setting of protective device |
| $I_{\Delta n}$ | = | rated residual operating current of the protective device in amperes |
| $I_t$ | = | tabulated current-carrying capacity of a cable |
| $I_z$ | = | current-carrying capacity of a cable for continuous service under the particular installation conditions concerned |
| $I^2t$ | = | energy let-through value of device |
| $I_2$ | = | current causing effective operation of the overload protective device |
| $k$ | = | material factor taken from Tables 43A and 54B to F |
| $k^2S^2$ | = | energy withstand of cable |
| $M$ | = | appropriate multiplier |
| $R_1$ | = | resistance of phase conductor from origin of installation to input terminal of the load |
| $R_2$ | = | resistance of protective conductor from the origin of installation to the earth terminal of the load |
| $S$ | = | conductor cross-sectional area |
| $t$ | = | time |
| $Z_e$ | = | that part of the earth fault loop impedance which is external to the installation |
| $Z_s$ | = | earth fault loop impedance |

N.B. Amendments to the Handbook arising from BS 7671 Amendment No 1 are identified in text by side bars and in charts and diagrams by underlining.

# B

## The Arrangement of Each Chapter

# PART 1
## OF THE REGULATIONS

# Scope, Object and Fundamental Requirements for Safety

This PART is divided into three CHAPTERS dealing in general terms with:

Chapter 11 deals with the scope of application of the Regulations and the types of work excluded.

Chapter 12 deals with the objectives to be achieved through compliance with the Regulations.

Chapter 13 deals with the fundamental requirements for safety encompassing workmanship, materials, conductors, joints, protective devices against overcurrent, earth leakage, and earth fault currents, the positioning of these devices and of the isolation and switching devices required to prevent danger.

Additionally Chapter 13 confirms the necessity for adequate safe access for operation and maintenance, suitability of the whole installation for the environment in which it has to operate, the requirements in regard to additions and alterations, and the need for initial and periodic testing of completed works.

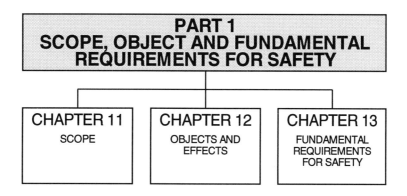

## Chapter 11 — Scope

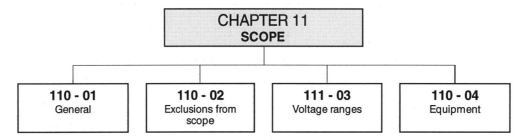

### Special Points to Note

(a)      The scope does not refer separately to permanent and temporary installations, the Regulations apply equally to both.

(b)      The Regulations for the additional and particular requirements for special installations or locations are defined in Section 6.

## Chapter 12 — Objects and Effects

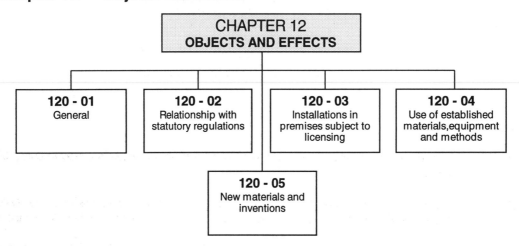

### Special Points to Note

(a)      The Regulations, although intended to be cited as a whole for contractual purposes, require to be supported by a detailed specification and, on installations of a difficult or special character, the advice of a suitably qualified electrical engineer.

(b)      The Regulations are not intended to instruct untrained persons.

(c)      Additionally, special requirements, not specifically detailed in the Regulations, apply to installations in premises subject to special licensing or statutory control, e.g. petrol pump/forecourt installations, quarry installations (Appendix 2 lists a number of such situations).

(d)      The Regulations include "property" in the protected categories as well as persons and livestock.

(e)      The Regulations provide a reminder that there is a need to ascertain and comply with licensing authority requirements.

(f)      The Regulations do not include use of new designs that lead to departures.

## Chapter 13 — Fundamental Requirements for Safety

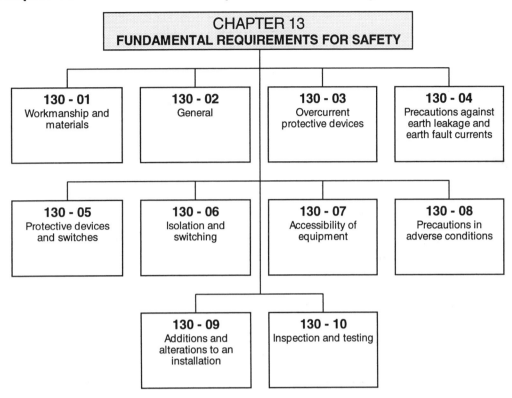

**Special Points to Note**

(a)     The requirement for good workmanship and proper materials is a Regulation (130 - 01 - 01).

(b)     It is a requirement (130 - 02 - 01) that future maintenance of the installation be taken into account, and the necessary provisions, such as sub-division for isolation, made.

(c)     Regulation 130 - 04 - 02 requires additions and alterations to an installation to comply with all relevant Regulations. Reference must also be made to Regulation 721 - 01 - 01 which imposes further requirements.

(d)     Regulation 130 - 10 - 01 emphasises the importance to be accorded to initial and periodic inspection and testing.

(e)     Some terms, e.g. overcurrent, residual current device, require reference to the Definitions. It is important that users are familiar with this terminology to avoid any misunderstanding over the meaning and use of the Regulations.

(f)     It should be noted that the requirments of Chapter 13 are not only those of the Wiring Regulations, but also in many instances are statutory, mainly under the Electricity Supply Regulations 1988 as amended, and the Electricity at Work Regulations 1989 (see also Appendix 2 of the Wiring Regulations).

# PART 2

## OF THE REGULATIONS

# Definitions

A thorough knowlege of the "Definitions" which form Part 2 of BS 7671: 1992 is essential to ensure common usage and understanding of the terminology in each clause. Care must be taken to ensure accurate usage of the terms by all concerned.

**Amendment No. 1 to BS 7671: 1992 – modifies the following definitions:**

Arm's Reach
Direct Contact
Distribution board
Earthing
Electrical Equipment
Emergency Stopping
Emergency Switching
Enclosure
Fixed Equipment
Functional Extra-Low Voltage (FELV)
Hot Air Sauna
Indirect Contact
Isolator
Protective Conductor (Fig. 2)
Separated Extra-Low Voltage (SELV)

**Amendment No. 1 to BS 7671 – incorporates the following new definitions:**

Class III Equipment
Consumer Unit
Disconnector
Flexible Cable
Flexible Cord
Functional Switching
Fuse Carrier
Harmonised Standard
Mobile Home
PELV (Protective Extra-Low Voltage)
Protective Multiple Earthing (PME)

Symbols used in the Regulations

**Figure B1: To Supplement the Definition of "System"**

**Important:** The meanings assigned to the various letters in the table below are for explanation only. Other combinations of the various letters to identify special systems is not permissible, unless these have received international approval.

| FIRST LETTER | SECOND LETTER | SUBSEQUENT LETTERS (if any) |
|---|---|---|
| **Earthing arrangement for source of energy** | **Relationship of exposed conductive parts of installation and earth** | **Arrangement of Neutral and Protective Conductors in TN Systems** |
| **T**<br>Direct connection of one or more points of the source of energy to earth | **N**<br>Direct connection of exposed conductive parts by protective conductors to earthed point of source of energy (usually the neutral point for AC systems) | **C**<br>Neutral and protective conductor combined in a single conductor |
| | | **S**<br>Separate conductors for neutral and protective functions |
| | | **C-S**<br>Neutral and protective functions combined in part of system only |
| **T**<br>Direct connection of one point to earth at source of energy | **T**<br>Direct electrical connection of the exposed conductive parts by protective conductors to earth via an earth electrode which is independent of the earthing arrangements of the source of energy | |
| **I**<br>All live parts of source of energy isolated from earth or one point earthed through an impedance | **T**<br>Direct electrical connection of the exposed conductive parts by protective conductors to earth via an earth electrode which is independent of the earthing arrangements of the source of energy | |

NOTES (i) An example of the use of a combined neutral and protective conductor (PEN Conductor) is earthed concentric wiring but special authorisation must be obtained from the appropriate authority.

(ii) Special requirements apply to installations supplied from PME systems (TN-C, TN-C-S) and the Supplier concerned should be consulted.

(iii) The 'Electricity Supply Regulations 1988 as amended', do not permit the use of IT systems for public supply.

(iv) Different systems may be encountered in the same building or location. For example, there may be extra low voltage circuits for a particular function and because of the nature of the source of supply (e.g. batteries) this will be a separate system from the general building system.

# PART 3

## OF THE REGULATIONS

# Assessment of General Characteristics

PART 3 is concerned with the aspects and purpose of every installation, and must be considered in detail before any design or installation is started.

The five chapters of PART 3 should not be regarded merely as a check-list, but rather as a step-by-step guide, each item of which must be examined carefully.

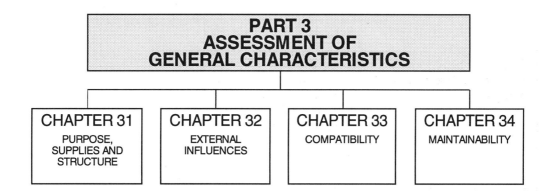

**Special Point to Note**

External influences – See also Regulations 512 - 06 - 01 and 02

## Chapter 31 — Purpose, Supplies and Structure

Study of this chapter will provide the governing parameters for assessing loads, the system (see Part 2), number type of conductors and the requirements for standby supplies together with earthing.

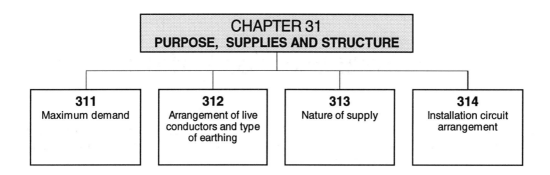

# Protection for Safety

Having considered the purpose for which the installation is being designed (PART 3), the designer and installer must concern themselves with the safety of the installation from the point of view of protecting the users, the property and the installation.

The hazards to be guarded against are electric shock, fire, burns, and the effects on the installation of overload current, fault current and undervoltage. The necessary measures must also be taken to ensure that the installation can be controlled properly and operated safely.

Chapters 41, 42, 43, 45 and 46 are concerned with the basic measures to be adopted for protection and Chapter 47 deals with their application. Reference is also necessary to PART 5 – Selection and Erection of Equipment – and to PART 7 – Inspection and Testing.

Whilst familiarity with PART 4 is essential, the presentation of the various protective measures as Topic Charts and Illustrations, later in this Handbook, is intended to draw together the various requirements for ease of reference.

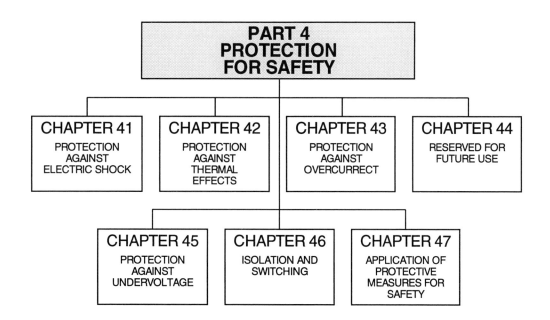

## Chapter 41 — Protection against Electric Shock

This chapter is concerned with protection of the users of the installation, whether they be concerned with its operation or availing themselves of its facilities. The provisions are qualified and to some extent explained by Chapter 47 and reference must be made to this chapter.

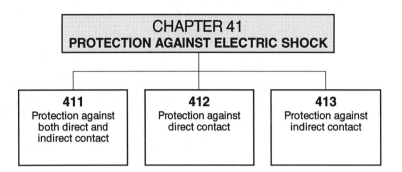

**Special Points to Note**

(a)     The requirement in Regulation 413 - 02 - 05 to allow for the change in conductor resistance due to temperature rise is deemed to be satisfied if the protective device complies with Appendix 3 and the circuit loop impedance is within the appropriate values in Regulation 413 - 02.

(b)     The range of $Z_S$ values and cpc impedances are derived from BS 3871 and BS EN 60898: 1991.

(c)     Loop impedance values in tables are related to normal operating temperatures. If the conductors are at a different temperature when tested an appropriate adjustment must be made.

(d)     When using Table 41E only phase to phase voltage is available if the neutral is not distributed, i.e. connected to the load.

## Chapter 42 — Protection against Thermal Effects

This chapter is concerned with the measures to be adopted to prevent danger from fire and burns which may be caused by the heat generated by fixed electrical equipment.

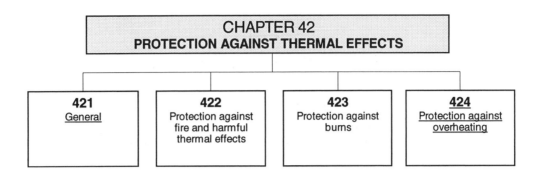

**Special Points to Note**

(a)     This chapter places an onus on the designer to decide on safe distances or materials to provide protection against thermal effects.

(b)     Regulation 422 - 01 - 04 limits the use of open back consumer units, in addition see Section 526 - 03.

(c)     Regulation 423 - 01 - 01 recognises that the same material in different situations can be allowed to reach different temperatures before being dangerous.

(d)     Regulation 423 - 01 - 01 states that certain fixed appliances, e.g. storage heaters, cannot be regarded as being sufficiently guarded to prevent accidental contact. Nevertheless, having regard to their purpose and if they comply with the appropriate British Standard, their use is admissible.

(e)     A requirement to ensure air flow is present when the heating element of a forced air heater is on, together with two temperature limiting devices is inherent in Regulation 424 - 01 - 01.

## Chapter 43 — Protection against Overcurrent

This chapter is mainly concerned with the factors which must be considered at the design stage of an installation. The chapter must be read in conjunction with other chapters (see Figure A3) and Topic Chart 5 – Overcurrent.

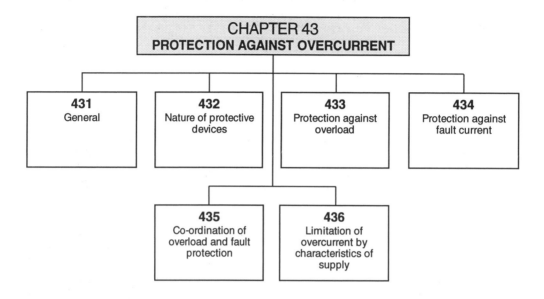

**Special Points to Note**

(a)     Regulation 433 - 02 - 04 sets out the deemed to comply with Regulation 433 - 02 - 01 requirements, for ring final circuits protected by a 30A or 32A protective device.

(b)     Regulation 434 - 04 - 01 requires a calculation to be made to check the short circuit withstand of conductors and cpc if the fault did not affect all parallel conductors thus possibly extending the disconnection time.

## Chapter 45 — Protection against Undervoltage

This chapter is concerned with the prevention of possible danger and/or damage caused by supply voltage reduction, or cessation and subsequent resumption of supply. It should be noted that Regulation 552 - 01 - 03 is also relevant.

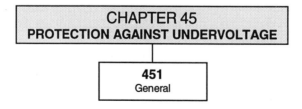

## Chapter 46 — Isolation and Switching

This chapter is concerned with the basic requirements to prevent or remove hazards associated with the electrical installation or electrical equipment and machines.

The chapter must be read in conjunction with other chapters (see Figure A3) and Topic Chart 4 Isolation and Switching.

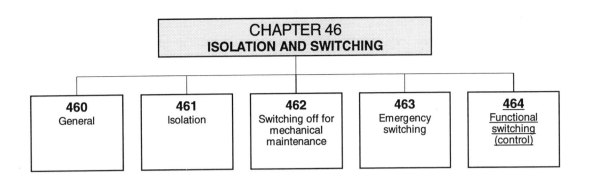

# CHAPTER 46
# ISOLATION AND SWITCHING

| 460 General | 461 Isolation | 462 Switching off for mechanical maintenance | 463 Emergency switching | 464 Functional switching (control) |

**Sections 461, 462 and 463**

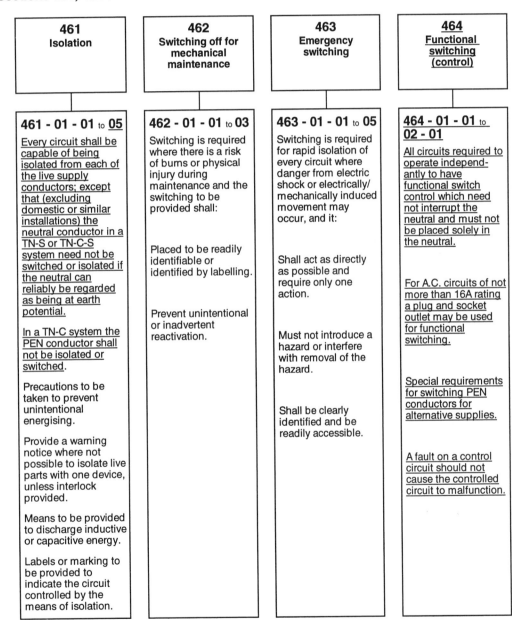

| 461 Isolation | 462 Switching off for mechanical maintenance | 463 Emergency switching | 464 Functional switching (control) |
|---|---|---|---|

**461 - 01 - 01 to 05**

Every circuit shall be capable of being isolated from each of the live supply conductors; except that (excluding domestic or similar installations) the neutral conductor in a TN-S or TN-C-S system need not be switched or isolated if the neutral can reliably be regarded as being at earth potential.

In a TN-C system the PEN conductor shall not be isolated or switched.

Precautions to be taken to prevent unintentional energising.

Provide a warning notice where not possible to isolate live parts with one device, unless interlock provided.

Means to be provided to discharge inductive or capacitive energy.

Labels or marking to be provided to indicate the circuit controlled by the means of isolation.

**462 - 01 - 01 to 03**

Switching is required where there is a risk of burns or physical injury during maintenance and the switching to be provided shall:

Placed to be readily identifiable or identified by labelling.

Prevent unintentional or inadvertent reactivation.

**463 - 01 - 01 to 05**

Switching is required for rapid isolation of every circuit where danger from electric shock or electrically/mechanically induced movement may occur, and it:

Shall act as directly as possible and require only one action.

Must not introduce a hazard or interfere with removal of the hazard.

Shall be clearly identified and be readily accessible.

**464 - 01 - 01 to 02 - 01**

All circuits required to operate independantly to have functional switch control which need not interrupt the neutral and must not be placed solely in the neutral.

For A.C. circuits of not more than 16A rating a plug and socket outlet may be used for functional switching.

Special requirements for switching PEN conductors for alternative supplies.

A fault on a control circuit should not cause the controlled circuit to malfunction.

**Special Points to Note**

(a)     The protective or PEN conductor shall not incorporate any means of isolation or switching (460 - 01 - 03) except as 460 - 01 - 05 and in a TN-C system shall not be isolated or switched.

(b)     Provision need not be made for isolation of the neutral conductor in TN-S and TN-C-S systems unless specifically required by 460 - 01 - 02.

## Chapter 47 — Application of Protective Measures for Safety

This chapter is extremely important for it is here that the qualification and, in some cases, amplification of the requirements of Chapters 41, 43 and 46 are contained.

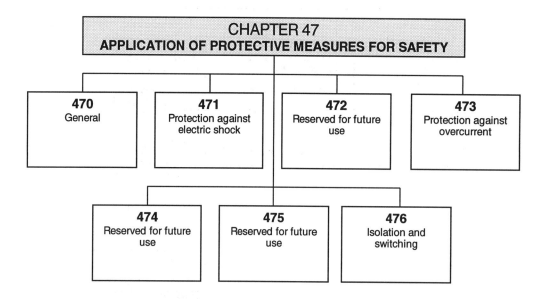

### Special Point to Note

No reference is made in Chapter 47 to undervoltage protection, which is dealt with in Chapter 45 and Regulation 552 - 01 - 03.

**Section 471**

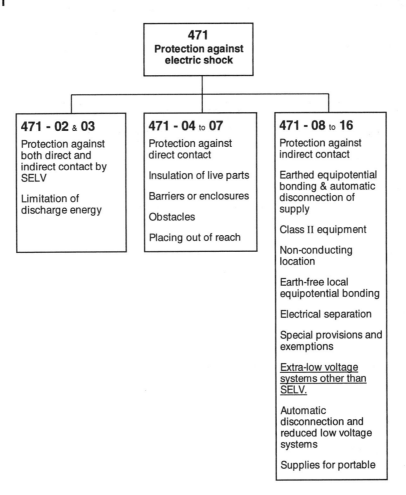

**Section 473**

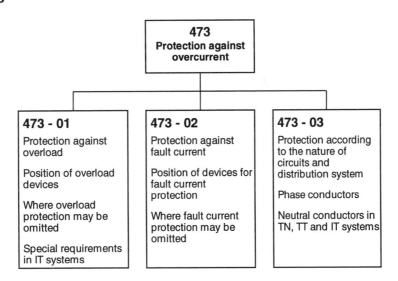

**Special Points to Note**

(a)    Regulation 473 - 01 - 04 now allows the Supplier's device to provide overload protection where the agreement of the Supplier is obtained.

(b)    Regulation 473 - 03 - 02 covers special requirements for TT systems.

**476**
**Isolation and switching**

---

**476 - 01**

General

Every circuit must have means of isolation and on-load switching

If the Supplier agrees, his equipment may be used for this purpose

Where one device serves more than one function, every requirement for each function must be satisfied

Groups of circuits may be switched by a common device

A main switch for use by an unskilled person must interrupt both live conductors of a single phase supply.

---

**476 - 02**

Isolation

Means of isolation to be provided at origin of installation

Interlocking required or location accessible to skilled persons only

Requirements where disconnectors are remote from equipment to be isolated

Every motor to have means of isolation

Special requirements for discharge lighting circuits operating above low voltage

---

**476 - 03**

Emergency switching

Every device to be readily accessible

Where danger could arise from incorrect use, emergency switches must be accessible to skilled persons only

Means of emergency switching on-load for every machine

Fireman's emergency switch for discharge lighting operating above low voltage

---

**PART 5**

OF THE REGULATIONS

# Selection and Erection of Equipment

This PART is concerned with the materials and methods of installation which are necessary to attain the standard of safety required by the Regulations.

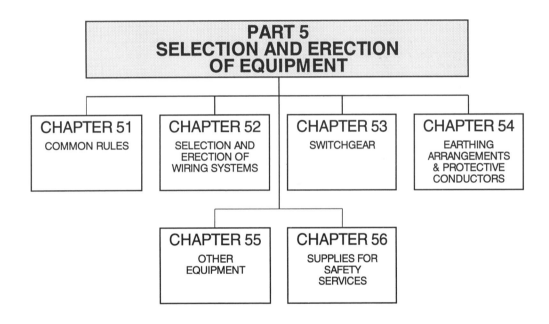

# Chapter 51 — Common Rules

In this chapter the rules that apply to all electrical installations, for any purpose or location, are set down as the basic requirements.

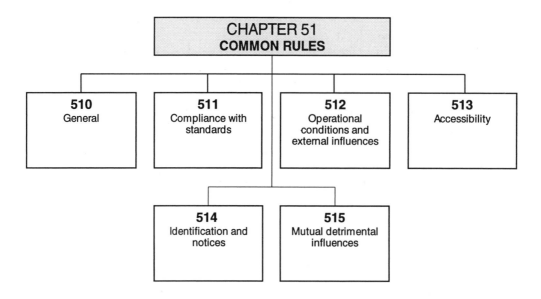

## Section 512

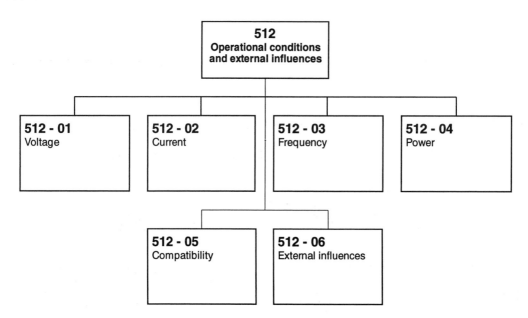

**Section 514**

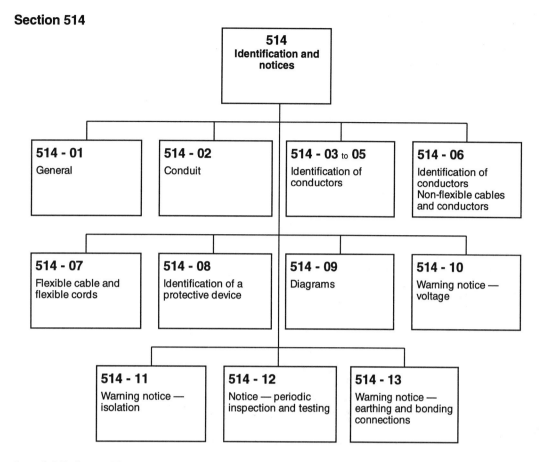

**Special Point to Note:**

Regulation 514 - 01 - 02 requires identification of wiring by arrangement or marking for inspection and testing etc.

## Chapter 52 — Selection and Erection of Wiring Systems

All installations need a wiring system and this chapter contains the requirements that govern the type of wiring system to be selected.

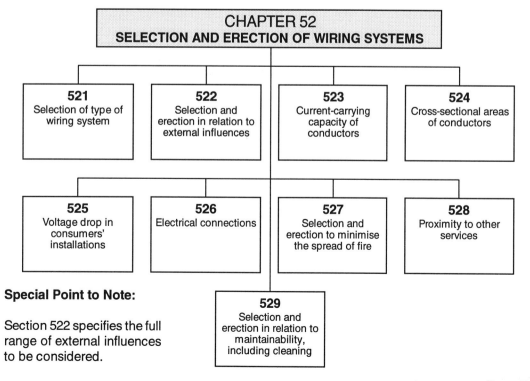

**Special Point to Note:**

Section 522 specifies the full range of external influences to be considered.

## Section 521

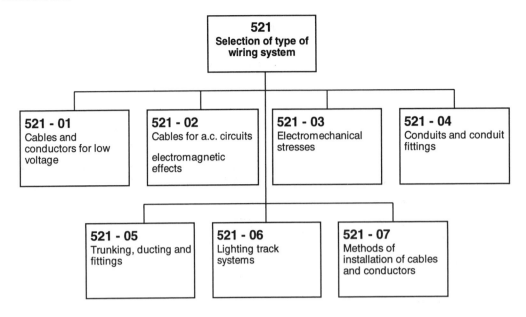

**521**
**Selection of type of wiring system**

**521 - 01**
Cables and conductors for low voltage

**521 - 02**
Cables for a.c. circuits electromagnetic effects

**521 - 03**
Electromechanical stresses

**521 - 04**
Conduits and conduit fittings

**521 - 05**
Trunking, ducting and fittings

**521 - 06**
Lighting track systems

**521 - 07**
Methods of installation of cables and conductors

## Section 522

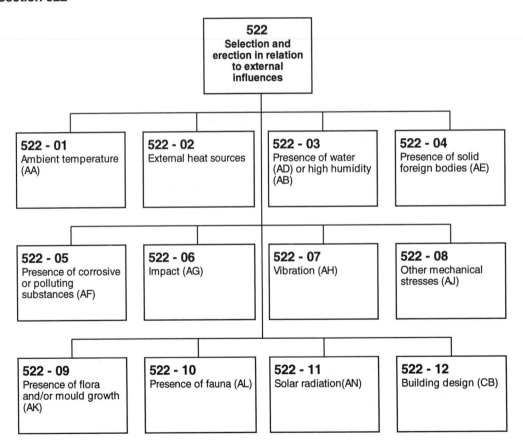

**522**
**Selection and erection in relation to external influences**

**522 - 01**
Ambient temperature (AA)

**522 - 02**
External heat sources

**522 - 03**
Presence of water (AD) or high humidity (AB)

**522 - 04**
Presence of solid foreign bodies (AE)

**522 - 05**
Presence of corrosive or polluting substances (AF)

**522 - 06**
Impact (AG)

**522 - 07**
Vibration (AH)

**522 - 08**
Other mechanical stresses (AJ)

**522 - 09**
Presence of flora and/or mould growth (AK)

**522 - 10**
Presence of fauna (AL)

**522 - 11**
Solar radiation(AN)

**522 - 12**
Building design (CB)

**Section 523**

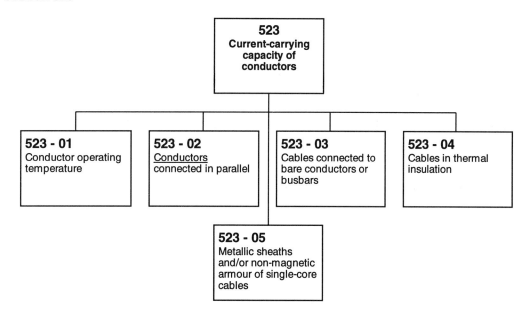

**Section 524**

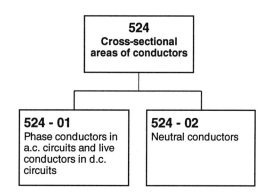

**Section 526**

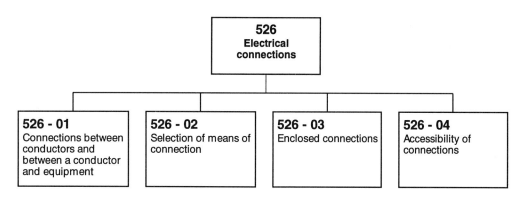

**Section 527**

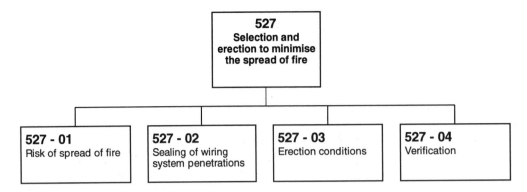

**Section 528**

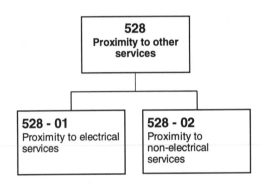

## Chapter 53 — Switchgear

Before a selection of switchgear can be made, many factors have to be assessed for their influence on the choice. These factors are set out in this chapter.

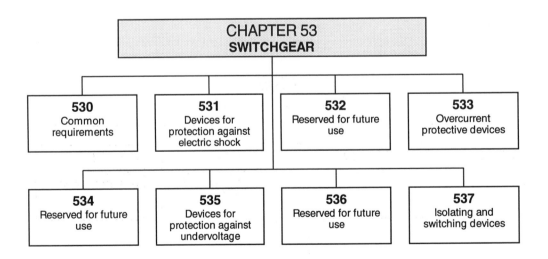

**Section 531**

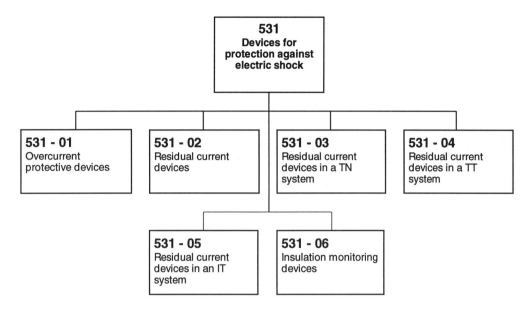

**Section 533**

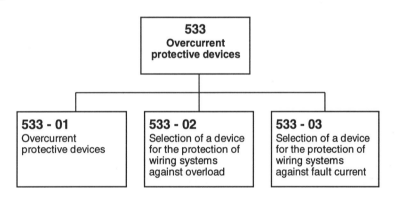

**Section 537**

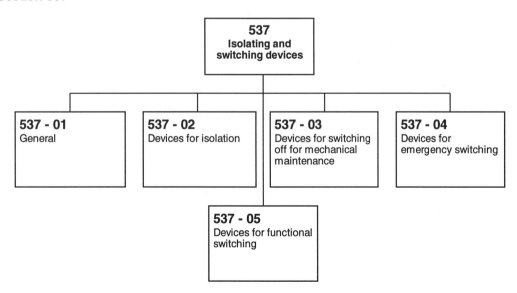

# Chapter 54 — Earthing Arrangements and Protective Conductors

In order to ensure that protective devices operate, specific earthing arrangements are required supplemented by protective conductors.  The applicable methods can be determined by the application of the Regulations in this chapter.

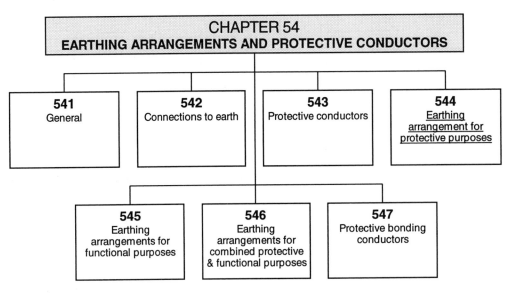

**Special Point to Note**

(a)      If a separate protective conductor is used it shall be installed in close proximity to the cable containing the live conductors (544 - 01 - 01)

**Section 542**

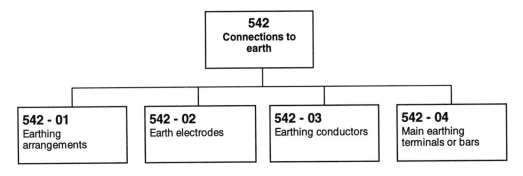

**Section 543**

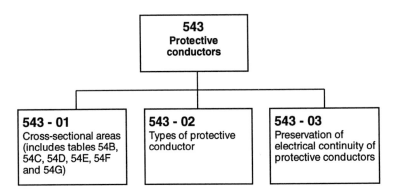

**Section 547**

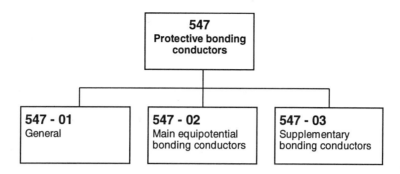

## Chapter 55 — Other Equipment

All items of equipment connected to an installation have to meet the requirements set out in this chapter.

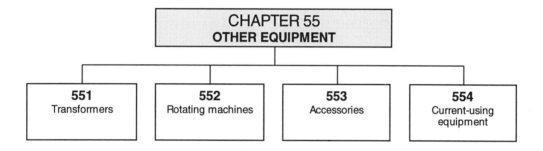

**Section 553**

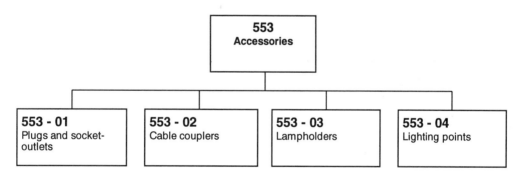

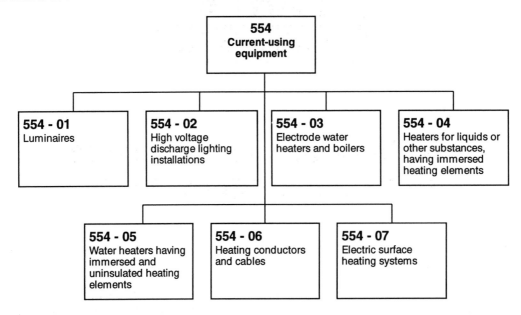

## Chapter 56 — Supplies for Safety Services

Within most non-domestic buildings there are circuits that must continue to operate in an emergency such as fire detection and alarm, smoke ventilation fans, sprinkler pumps, etc. The supplies to these safety services need to comply with the special requirements of this chapter to ensure they remain working for as long as possible.

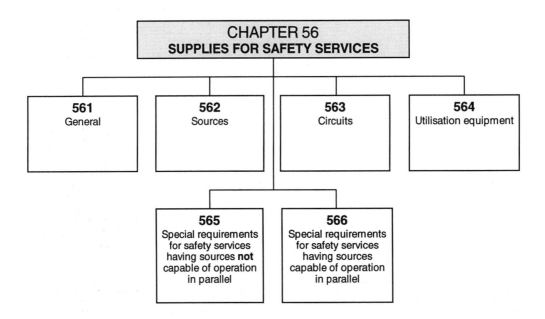

**Special Points to Note**

(a)     Equipment to be located to facilitate inspection, testing and maintenance.

(b)     Any fumes from a source must not penetrate occupied areas.

(c)     Regulation 566-01-02 draws attention to circulating currents caused by harmonics.

# Special Installations or Locations – Particular Requirements

Parts 1 - 5 give the designer or installer all the requirements for installations in standard locations either Commercial, Industrial or Domestic.

However, there are special locations either within a standard installation or existing in their own right. These special installations or locations by their nature or the nature of the activity that takes place within them, need special attention.

This part of the Regulations supplements or modifies the requirements of other sections of the Regulations.

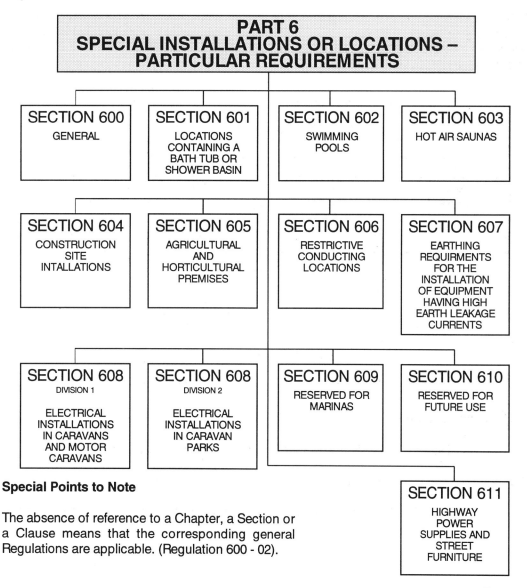

**Special Points to Note**

The absence of reference to a Chapter, a Section or a Clause means that the corresponding general Regulations are applicable. (Regulation 600 - 02).

**Section 601**

A room containing a bath tub or shower basin is there for the express purpose of allowing people to be wet, thereby reducing their body resistance. To minimise the possibility of electric shock under these conditions, the requirements of this chapter must be incorporated in the installation.

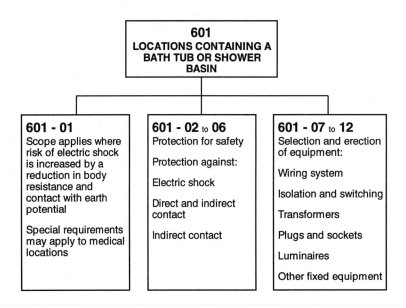

**601**
**LOCATIONS CONTAINING A BATH TUB OR SHOWER BASIN**

**601 - 01**
Scope applies where risk of electric shock is increased by a reduction in body resistance and contact with earth potential

Special requirements may apply to medical locations

**601 - 02 to 06**
Protection for safety

Protection against:

Electric shock

Direct and indirect contact

Indirect contact

**601 - 07 to 12**
Selection and erection of equipment:

Wiring system

Isolation and switching

Transformers

Plugs and sockets

Luminaires

Other fixed equipment

**Section 602**

The comments of Section 601 apply to this chapter with the additional risk of persons using electrical equipment on the poolside or in the water.

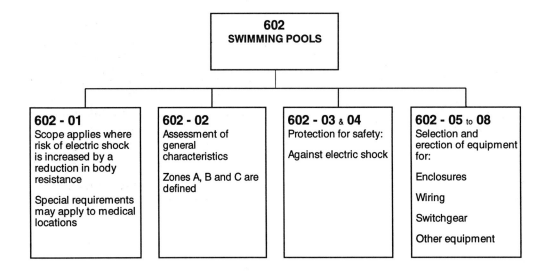

**602**
**SWIMMING POOLS**

**602 - 01**
Scope applies where risk of electric shock is increased by a reduction in body resistance

Special requirements may apply to medical locations

**602 - 02**
Assessment of general characteristics

Zones A, B and C are defined

**602 - 03 & 04**
Protection for safety:

Against electric shock

**602 - 05 to 08**
Selection and erection of equipment for:

Enclosures

Wiring

Switchgear

Other equipment

**Special Point to Note**

(a)     Regulation 602 - 03 - 02 no longer requires a metal grid in a solid floor of zones B and C. However, where one is provided it shall be supplementary bonded.

**Section 603**

The conditions in a sauna steam room pose similar problems to the two previous situations with the added hazard of periods of high humidity which can cause deterioration of the installation with consequent increase in risks.

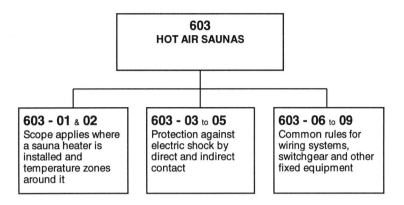

**603**
**HOT AIR SAUNAS**

**603 - 01 & 02**
Scope applies where a sauna heater is installed and temperature zones around it

**603 - 03 to 05**
Protection against electric shock by direct and indirect contact

**603 - 06 to 09**
Common rules for wiring systems, switchgear and other fixed equipment

**Section 604**

Because the installations on a construction site are temporary they may be inherently less well protected and more exposed to damage during building operations. To give site operatives further protection from danger, this chapter sets out special requirements.

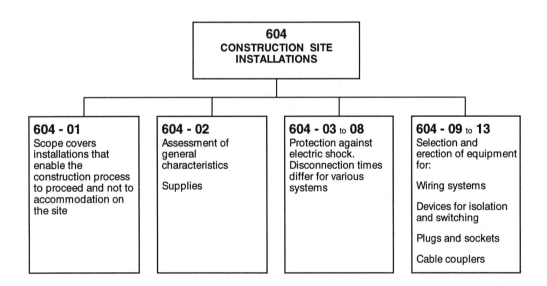

**604**
**CONSTRUCTION SITE INSTALLATIONS**

**604 - 01**
Scope covers installations that enable the construction process to proceed and not to accommodation on the site

**604 - 02**
Assessment of general characteristics

Supplies

**604 - 03 to 08**
Protection against electric shock. Disconnection times differ for various systems

**604 - 09 to 13**
Selection and erection of equipment for:

Wiring systems

Devices for isolation and switching

Plugs and sockets

Cable couplers

**Section 605**

Many installations in and around agricultural and horticultural buildings are in the open or virtually so. Also many of them are in areas where livestock are normally located. This section sets out the requirements to minimise the risk to persons and livestock.

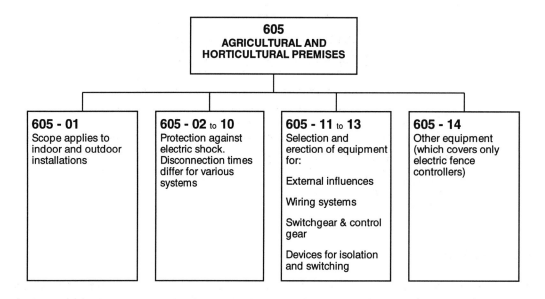

**Section 606**

Occasionally an operative is working in close proximity to conducting materials which restrict the person's movement, e.g. inside a metal vessel. Since the operative cannot move to avoid danger, this section sets down specific requirements to minimise the additional dangers arising from the location.

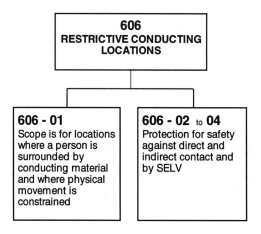

**Section 607**

Much of the electronic equipment now in use has an inherently high earth leakage current. To maintain safety under these conditions the earthing requirements of this chapter must be adhered to.

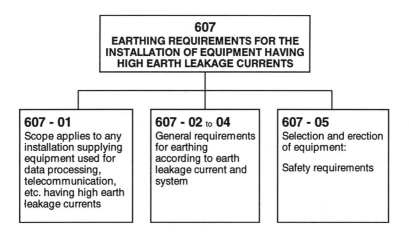

**Section 608  Division One**

Whilst this type of accommodation is basically like a house it is fed from a plug-in system and part of it may be subjected to a greater range of external influences. Thus additional requirements are set out in this chapter.

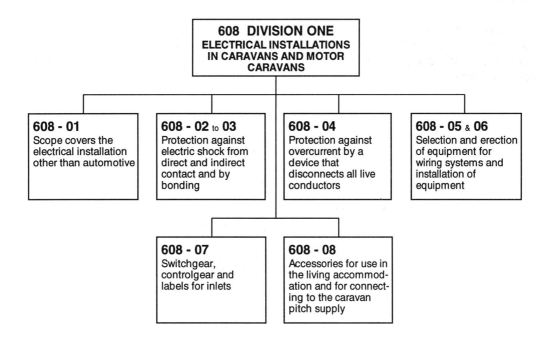

## Section 608  Division Two

Since the installation in a caravan park may supply installations, which are inferior or have been damaged during transit, this section sets out additional requirements to ensure the safety of the occupiers and restrict the influence on other circuits of a fault connected to one outlet.

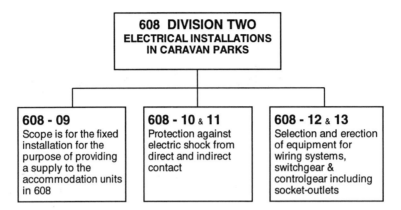

## Section 611

This section details additional requirements to minimise the possibility of electric shock to persons using public places, other than buildings, where there are electrical installations.

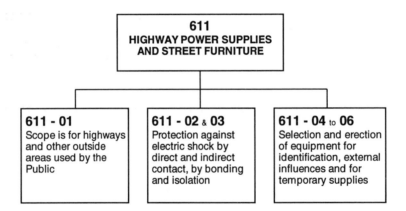

# PART 7

## OF THE REGULATIONS

# Inspection and Testing

After the design and the installation has been completed, there is a need to ensure that it meets the letter and the spirit of the Regulations. This is obtained by inspection, testing and recording as set out in this part.

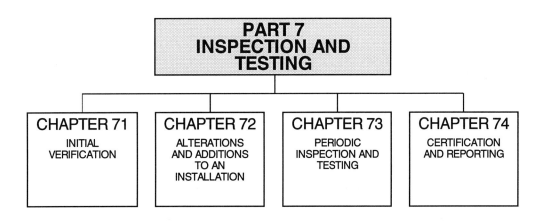

# Chapter 71 — Initial Verification

Within this chapter are the requirements to ensure the new installation is in a suitable state to be put into service.

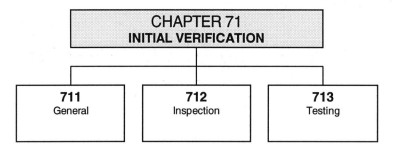

## Sections 711 and 712

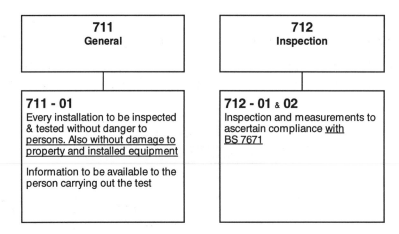

## Section 713

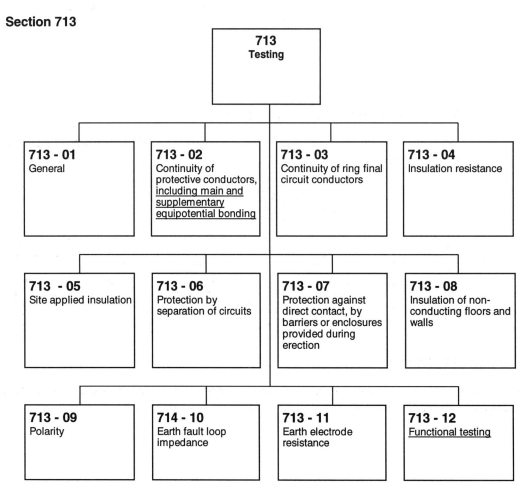

## Chapter 72 — Alterations and Additions to Installations

A very short chapter containing two regulations, one with immense power in that it implies a need to check the safety of the existing installation after any further work (but not necessarily to the latest BS7671 requirements).

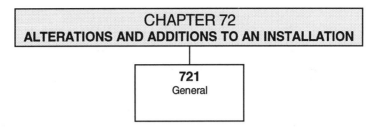

## Chapter 73 — Periodic Inspection and Testing

Whilst the responsibility for initiating this is invested in the owner or user of the installation, the work will have to be carried out by a competent person. This chapter sets out the methods and requirements.

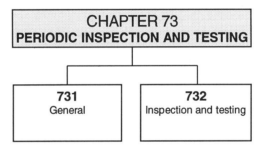

## Chapter 74 — Certification and Reporting

A requirement for uniformity of information about installations and the applicable test results ensures that persons carrying out the tests adopt a consistent and professional approach.

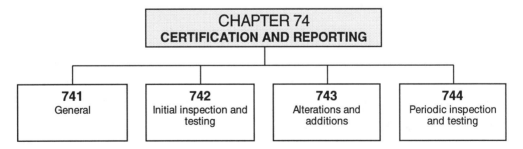

**Special Points to Note**

Regulation 741 - 01 - 01 requires a statement in respect of the design. Care must be taken before signing this if the design has not been carried out by the installer.

# C

## Topic Charts for Decision-making and Guidance

# Topic Chart 1

# Assessment of General Characteristics

## Purposes, Supplies and Structure

The Maximum Demand (M.D.), of any installation, is the summation of all the installed loads less an estimation of the reduction in load due to diversity. Diversity is assessed from a knowledge of the operation of the system to determine what installed loads will not be on full load at the same time due to such factors as automatic time or temperature control, occupation patterns etc.

M.D. can be calculated on two bases:-

    half hour for transformers, cables etc. and
    short time for overload current, i.e. fuse or circuit breaker rating

For smaller installations guidance is given in IEE Guidance Notes, however, it must be recognized that whilst tables are all very well, there is no substitute for experience. Tables should be used with care and within the limits of one's own competence. For larger installations the assessment of M.D. and the diversity must be undertaken by a competent person.

Regulations in Sections 312 and 313 are intended to direct the designer (and in this context, the installer may also be acting as the designer) towards a consideration of the nature and characteristics of the supply on which the requirements for safety contained in Part 4 are selected. Examples of the information which will be required are the design levels for the prospective short circuit current, and the method of earthing which will affect the automatic disconnection times specified in Part 4.

Regulations in Section 314 contains the basic requirements for sub-division of the installation into manageable parts in order that any danger is avoided and inconvenience is minimized in the event of a fault, and to allow normal maintenance procedures to be carried out safely.

## External Influences

No specific guidance is given on External Influences because the international work on this subject is still under consideration. It is important to ensure, however, that every item of equipment is suitable for the environment in which it is installed. Reference should also be made to Appendix 5 of BS 7671: 1992 which contains guidance on the factors to be considered when selecting equipment for particular locations.

## Compatibility

The list of characteristics given later is not exhaustive and is only offered as a guide. The designer, installer, and the inspection and test engineer should ensure that all these aspects are taken into consideration.

The growing use of electronic equipment in general electrical installations (e.g. dimmer switches) merits particular attention, because in certain circumstances it can give rise to harmonics and peak voltages which can affect other equipment (see note on Page 83).

# TOPIC CHART 1 (Information) Assessment of General Characteristics

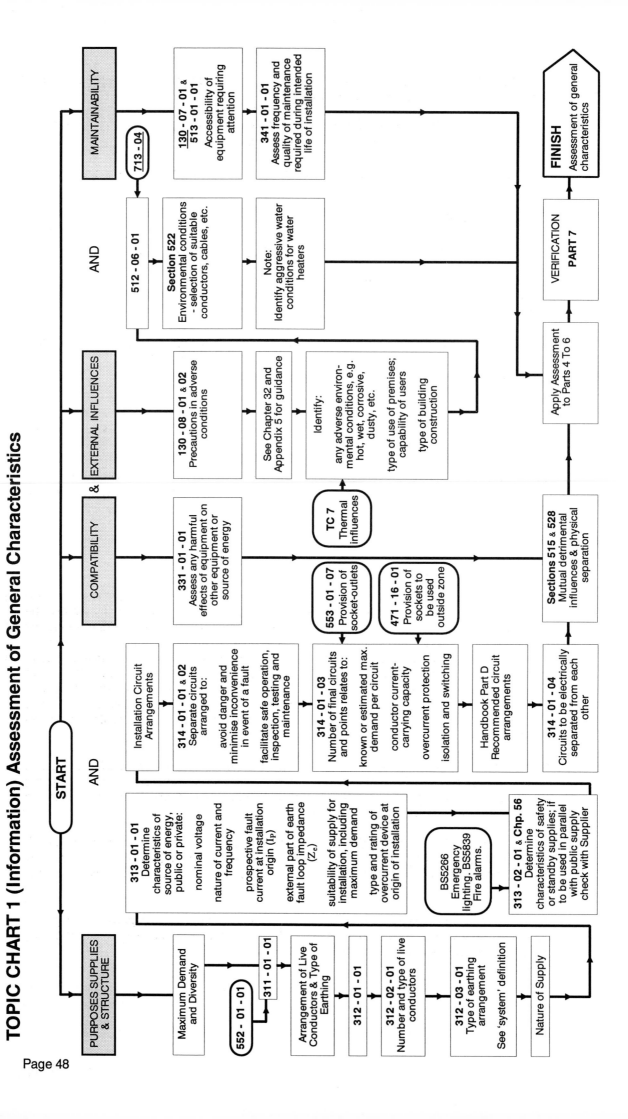

**START**

**PURPOSES SUPPLIES & STRUCTURE**

Maximum Demand and Diversity

**552 - 01 - 01**

**311 - 01 - 01**

Arrangement of Live Conductors & Type of Earthing

**312 - 01 - 01**

**312 - 02 - 01**
Number and type of live conductors

**312 - 03 - 01**
Type of earthing arrangement

See 'system' definition

Nature of Supply

**313 - 01 - 01**
Determine characteristics of source of energy, public or private:

- nominal voltage
- nature of current and frequency
- prospective fault current at installation origin ($I_p$)
- external part of earth fault loop impedance ($Z_e$)
- suitability of supply for installation, including maximum demand
- type and rating of overcurrent device at origin of installation

**BS5266** Emergency lighting. BS5839 Fire alarms.

**313 - 02 - 01 & Chp. 56**
Determine characteristics of safety or standby supplies; if to be used in parallel with public supply check with Supplier

**AND**

Installation Circuit Arrangements

**314 - 01 - 01 & 02**
Separate circuits arranged to:

- avoid danger and minimise inconvenience in event of a fault
- facilitate safe operation, inspection, testing and maintenance

**314 - 01 - 03**
Number of final circuits and points relates to:

- known or estimated max. demand per circuit
- conductor current-carrying capacity
- overcurrent protection
- isolation and switching

Handbook Part D Recommended circuit arrangements

**314 - 01 - 04**
Circuits to be electrically separated from each other

**553 - 01 - 07** Provision of socket-outlets

**471 - 16 - 01** Provision of sockets to be used outside zone

**COMPATIBILITY**

**331 - 01 - 01**
Assess any harmful effects of equipment on other equipment or source of energy

**Sections 515 & 528**
Mutual detrimental influences and physical separation

**& EXTERNAL INFLUENCES**

**130 - 08 - 01 & 02**
Precautions in adverse conditions

See Chapter 32 and Appendix 5 for guidance

Identify:

- any adverse environmental conditions, e.g. hot, wet, corrosive, dusty, etc.
- type of use of premises; capability of users
- type of building construction

**TC 7** Thermal influences

**AND**

**512 - 06 - 01**

**Section 522**
Environmental conditions - selection of suitable conductors, cables, etc.

Note:
Identify aggressive water conditions for water heaters

**MAINTAINABILITY**

**713 - 04**

**130 - 07 - 01 & 513 - 01 - 01**
Accessibility of equipment requiring attention

**341 - 01 - 01**
Assess frequency and quality of maintenance required during intended life of installation

Apply Assessment to Parts 4 To 6

VERIFICATION
**PART 7**

**FINISH**
Assessment of general characteristics

In addition, interference may be caused by transient disturbances generated by switching of any appliance containing capacitance or inductance.

ERA Report 75-31 (Code of Practice for the Avoidance of Electrical Interference in Electronic Instrumentation and Systems) gives useful information in this area.

The following list contains some of the characteristics that affect other equipment or the supply.

1. Transient overvoltages
2. Rapidly fluctuating loads
3. Starting currents
4. Harmonic currents (such as with fluorescent lighting loads and variable speed drives)
5. Induction
6. DC component in AC waveform
7. Earth leakage currents
8. Any need for additional connections to an independent earth.

Increasing use of electronic equipment and controls makes the assessment of the harmful effect on the supply worthy of the fullest investigation and if in any doubt the situation should be discussed with the Supplier.

## Maintainability

Chapter 34 draws attention to the necessity of considering the use or abuse to which the installation will be subjected during its life.

As well as the reliability and durability of the installation techniques, materials and equipment, the accessibility of all equipment must be considered so that regular inspection and maintenance can be readily and safely carried out.

## Minimizing Danger and Inconvenience

Regulation 314 - 01 - 01 requires that circuits be arranged to avoid danger and minimize inconvenience in the event of a fault, and 314 - 01 - 02 requires consideration to be given to the consequences caused by the operation of a single protective device. The foreseeable dangers are more physical than electrical, and proper consideration of the difficulties which may arise from a faulty circuit will probably provide equally well against inconvenience as well as danger. It is fairly obvious that no matter how few lights are included in an installation, they should not all be on the same circuit, and a little thought will divide them to the best advantage e.g. in a small house with a staircase, the hall and landing should be on separate circuits, and it may be that a vertical rather than a horizontal division will provide some light in all parts of the premises with only one circuit operative.

The provision of an rcd in place of an isolating switch to control the whole installation will lead to inconvenience and perhaps danger in the event of a fault. It is preferable to omit the lighting circuits from the protection of the rcd, bearing in mind that the shock danger from phase to neutral at a lampholder is not protected by the rcd.

Alternatively, a split consumer unit may be used so that a common isolator provides a single main control to separate parts of the consumer unit. This allows each part of the consumer unit to have its own rcd or rcds. Thus the lighting circuit(s) can have rcd protection independent of any other circuit(s).

Consideration should also be given to "essential" circuits, such as those supplying deep-freeze units, in particular whether they should share rcd protection with other circuits, or whether they should be provided with their own rcd protection.

**Figure C1**

# Earthing Arrangements and Network Terminations Relating to the Various Systems

N.B. Systems IT and TN-C are not envisaged for general use and are, therefore, not shown.

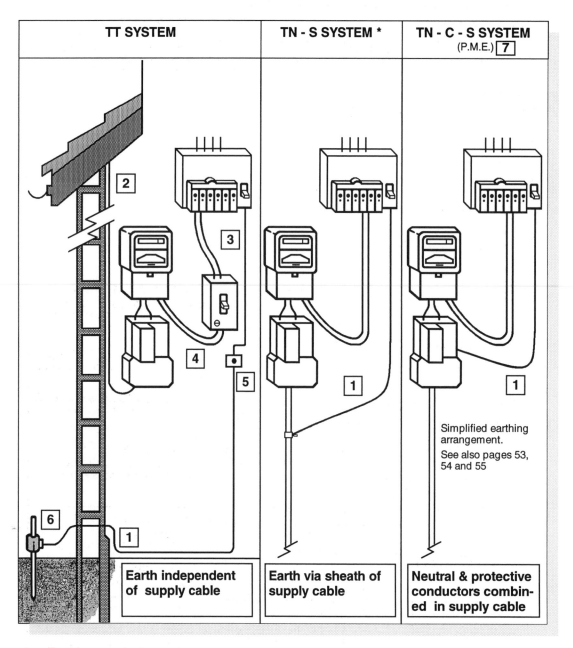

| TT SYSTEM | TN - S SYSTEM * | TN - C - S SYSTEM (P.M.E.) 7 |
|---|---|---|
| Earth independent of supply cable | Earth via sheath of supply cable | Neutral & protective conductors combined in supply cable |

Simplified earthing arrangement.
See also pages 53, 54 and 55

1   Earthing conductor
2   Cable from overhead supply
3   Earthing conductor connects directly to main earthing terminal
4   Cables not to be enclosed in metal conduit nor metal trunking as rcd does not protect against faults on supply side
5   Earthing conductor disconnecting terminal (desirable where the installation has its own earth electrodes)
6   Earth electrode  - see Fig. C6(c) page 000
7   Where a PME service is, or may be, involved, there should be early discussion and agreement with the appropriate Supplier, most of which have produced explanatory booklets showing their requirements for such installations. (See definition of System in Part 2 of BS 7671: 1992) (rcd may be used for supplementary protection against direct contact).

# Chapter 52
## Environmental Conditions
### Presence of water, dust, corrosive conditions

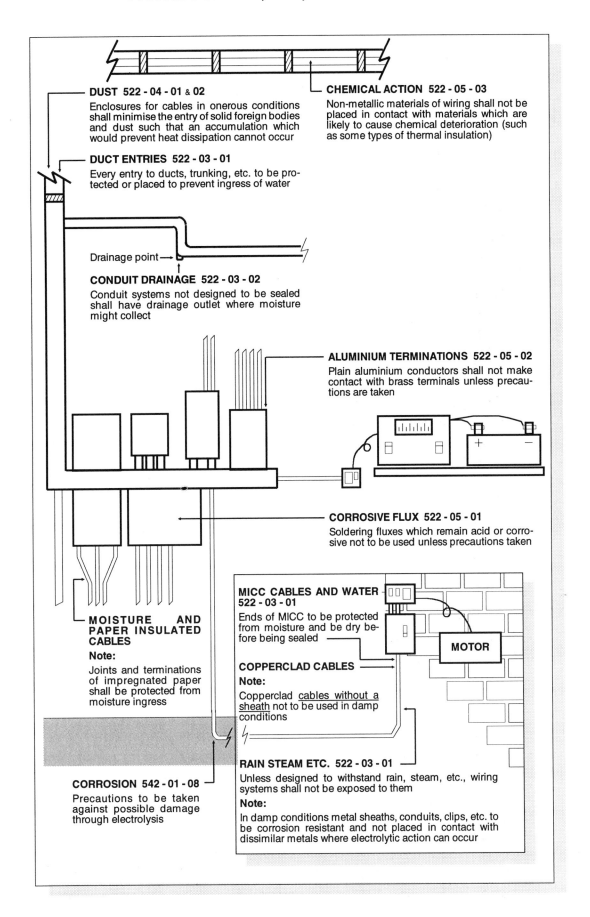

**DUST 522 - 04 - 01 & 02**

Enclosures for cables in onerous conditions shall minimise the entry of solid foreign bodies and dust such that an accumulation which would prevent heat dissipation cannot occur

**DUCT ENTRIES 522 - 03 - 01**

Every entry to ducts, trunking, etc. to be protected or placed to prevent ingress of water

Drainage point

**CONDUIT DRAINAGE 522 - 03 - 02**

Conduit systems not designed to be sealed shall have drainage outlet where moisture might collect

**CHEMICAL ACTION 522 - 05 - 03**

Non-metallic materials of wiring shall not be placed in contact with materials which are likely to cause chemical deterioration (such as some types of thermal insulation)

**ALUMINIUM TERMINATIONS 522 - 05 - 02**

Plain aluminium conductors shall not make contact with brass terminals unless precautions are taken

**CORROSIVE FLUX 522 - 05 - 01**

Soldering fluxes which remain acid or corrosive not to be used unless precautions taken

**MOISTURE AND PAPER INSULATED CABLES**

**Note:**

Joints and terminations of impregnated paper shall be protected from moisture ingress

**CORROSION 542 - 01 - 08**

Precautions to be taken against possible damage through electrolysis

**MICC CABLES AND WATER 522 - 03 - 01**

Ends of MICC to be protected from moisture and be dry before being sealed

**COPPERCLAD CABLES**

**Note:**

Copperclad cables without a sheath not to be used in damp conditions

**MOTOR**

**RAIN STEAM ETC. 522 - 03 - 01**

Unless designed to withstand rain, steam, etc., wiring systems shall not be exposed to them

**Note:**

In damp conditions metal sheaths, conduits, clips, etc. to be corrosion resistant and not placed in contact with dissimilar metals where electrolytic action can occur

**Figure C3 (a)**

# Chapter 52
## Environmental Conditions
### Ambient temperature and mechanical stresses
(See Figures C12(a) and (b) for mechanical supports)

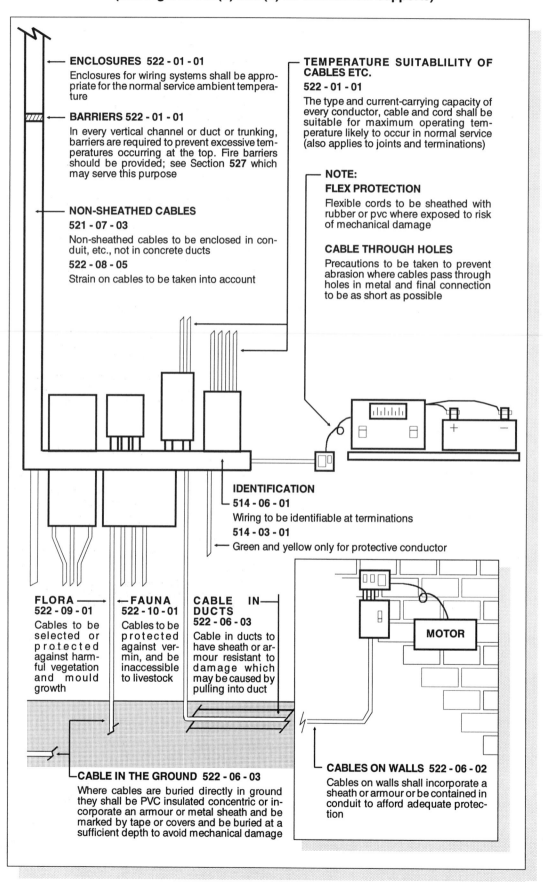

**ENCLOSURES 522 - 01 - 01**

Enclosures for wiring systems shall be appropriate for the normal service ambient temperature

**BARRIERS 522 - 01 - 01**

In every vertical channel or duct or trunking, barriers are required to prevent excessive temperatures occurring at the top. Fire barriers should be provided; see Section **527** which may serve this purpose

**NON-SHEATHED CABLES**
**521 - 07 - 03**

Non-sheathed cables to be enclosed in conduit, etc., not in concrete ducts

**522 - 08 - 05**

Strain on cables to be taken into account

**TEMPERATURE SUITABLLILITY OF CABLES ETC.**
**522 - 01 - 01**

The type and current-carrying capacity of every conductor, cable and cord shall be suitable for maximum operating temperature likely to occur in normal service (also applies to joints and terminations)

**NOTE:**

**FLEX PROTECTION**

Flexible cords to be sheathed with rubber or pvc where exposed to risk of mechanical damage

**CABLE THROUGH HOLES**

Precautions to be taken to prevent abrasion where cables pass through holes in metal and final connection to be as short as possible

**IDENTIFICATION**
**514 - 06 - 01**

Wiring to be identifiable at terminations

**514 - 03 - 01**

Green and yellow only for protective conductor

**FLORA**
**522 - 09 - 01**

Cables to be selected or protected against harmful vegetation and mould growth

**FAUNA**
**522 - 10 - 01**

Cables to be protected against vermin, and be inaccessible to livestock

**CABLE IN DUCTS**
**522 - 06 - 03**

Cable in ducts to have sheath or armour resistant to damage which may be caused by pulling into duct

**MOTOR**

**CABLE IN THE GROUND 522 - 06 - 03**

Where cables are buried directly in ground they shall be PVC insulated concentric or incorporate an armour or metal sheath and be marked by tape or covers and be buried at a sufficient depth to avoid mechanical damage

**CABLES ON WALLS 522 - 06 - 02**

Cables on walls shall incorporate a sheath or armour or be contained in conduit to afford adequate protection

# Chapter 52
## Environmental Conditions
### Ambient temperature and mechanical stresses (continued)
#### (See Figures C12(a) and (b) for mechanical supports)

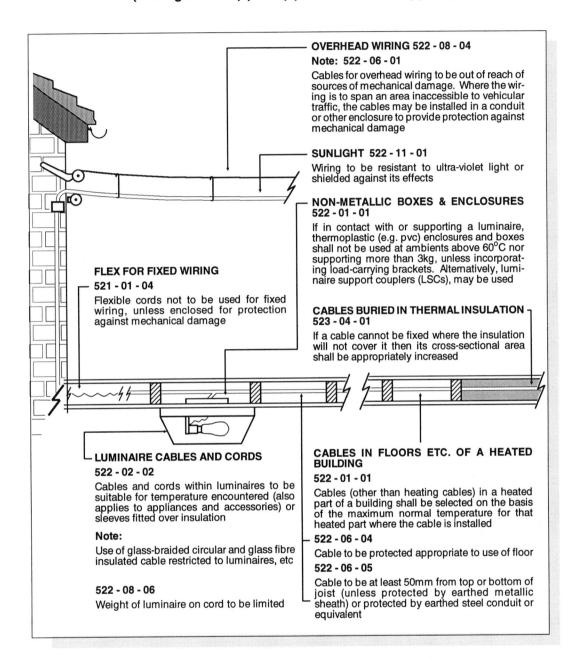

**OVERHEAD WIRING 522 - 08 - 04**

**Note: 522 - 06 - 01**

Cables for overhead wiring to be out of reach of sources of mechanical damage. Where the wiring is to span an area inaccessible to vehicular traffic, the cables may be installed in a conduit or other enclosure to provide protection against mechanical damage

**SUNLIGHT 522 - 11 - 01**

Wiring to be resistant to ultra-violet light or shielded against its effects

**NON-METALLIC BOXES & ENCLOSURES 522 - 01 - 01**

If in contact with or supporting a luminaire, thermoplastic (e.g. pvc) enclosures and boxes shall not be used at ambients above 60°C nor supporting more than 3kg, unless incorporating load-carrying brackets. Alternatively, luminaire support couplers (LSCs), may be used

**CABLES BURIED IN THERMAL INSULATION 523 - 04 - 01**

If a cable cannot be fixed where the insulation will not cover it then its cross-sectional area shall be appropriately increased

**FLEX FOR FIXED WIRING**

**521 - 01 - 04**

Flexible cords not to be used for fixed wiring, unless enclosed for protection against mechanical damage

**LUMINAIRE CABLES AND CORDS**

**522 - 02 - 02**

Cables and cords within luminaires to be suitable for temperature encountered (also applies to appliances and accessories) or sleeves fitted over insulation

**Note:**

Use of glass-braided circular and glass fibre insulated cable restricted to luminaires, etc

**522 - 08 - 06**

Weight of luminaire on cord to be limited

**CABLES IN FLOORS ETC. OF A HEATED BUILDING**

**522 - 01 - 01**

Cables (other than heating cables) in a heated part of a building shall be selected on the basis of the maximum normal temperature for that heated part where the cable is installed

**522 - 06 - 04**

Cable to be protected appropriate to use of floor

**522 - 06 - 05**

Cable to be at least 50mm from top or bottom of joist (unless protected by earthed metallic sheath) or protected by earthed steel conduit or equivalent

# S.E.L.V. Circuits

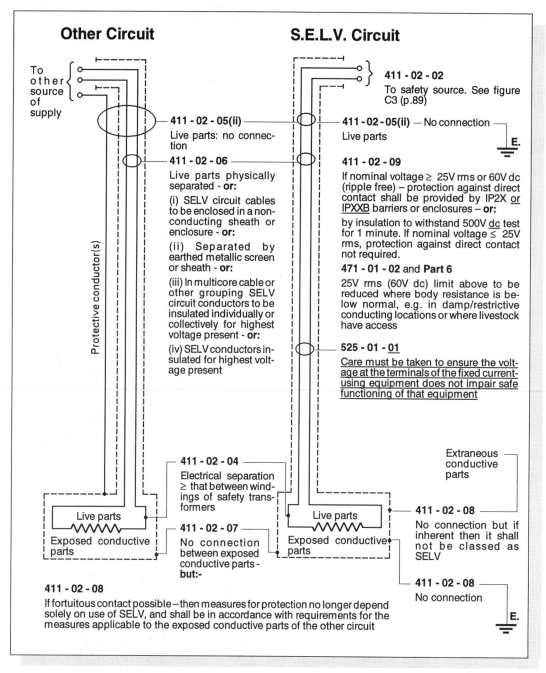

## Other Circuit

To other source of supply

Protective conductor(s)

**411 - 02 - 05(ii)**
Live parts: no connection

**411 - 02 - 06**
Live parts physically separated - **or:**

(i) SELV circuit cables to be enclosed in a non-conducting sheath or enclosure - **or:**

(ii) Separated by earthed metallic screen or sheath - **or:**

(iii) In multicore cable or other grouping SELV circuit conductors to be insulated individually or collectively for highest voltage present - **or:**

(iv) SELV conductors insulated for highest voltage present

**411 - 02 - 04**
Electrical separation ≥ that between windings of safety transformers

**411 - 02 - 07**
No connection between exposed conductive parts - **but:-**

Live parts

Exposed conductive parts

**411 - 02 - 08**
If fortuitous contact possible – then measures for protection no longer depend solely on use of SELV, and shall be in accordance with requirements for the measures applicable to the exposed conductive parts of the other circuit

## S.E.L.V. Circuit

**411 - 02 - 02**
To safety source. See figure C3 (p.89)

**411 - 02 - 05(ii)** — No connection —
Live parts
E.

**411 - 02 - 09**
If nominal voltage ≥ 25V rms or 60V dc (ripple free) – protection against direct contact shall be provided by IP2X or IPXXB barriers or enclosures – **or:**

by insulation to withstand 500V dc test for 1 minute. If nominal voltage ≤ 25V rms, protection against direct contact not required.

**471 - 01 - 02** and **Part 6**
25V rms (60V dc) limit above to be reduced where body resistance is below normal, e.g. in damp/restrictive conducting locations or where livestock have access

**525 - 01 - 01**
Care must be taken to ensure the voltage at the terminals of the fixed current-using equipment does not impair safe functioning of that equipment

Extraneous conductive parts

Live parts

Exposed conductive parts

**411 - 02 - 08**
No connection but if inherent then it shall not be classed as SELV

**411 - 02 - 08**
No connection
E.

| ITEM | 411 - 02 - 10 Plugs and sockets for SELV circuits | | For comparison, plugs and sockets for Extra-Low Voltage systems other than SELV. 471 - 14 - 01 to 06 | | |
|---|---|---|---|---|---|
| **SOCKET OUTLETS** | Shall exclude plugs of other voltage systems in use in same premises | Shall not have protective conductor connection **except** | If no protective conductor connection | **471 - 14 - 06** Shall exclude plugs of other voltage systems in use in the same premises | **Note:** Protective conductor connection required |
| **PLUGS** | **411 - 02 - 10** Shall not enter a s/o of other voltage systems in use in same premises | Luminaire supporting couplers permitted subject to 471 - 14 - 06 | | **471 - 14 - 06** Shall not enter a s/o of other voltage systems in use in same premises | Luminaire supporting couplers permitted subject to 471 - 14 - 06 |

**Note:**

If all the requirements for SELV circuits cannot be met, e.g. an earth connection is required to the circuit, then the system becomes PELV and no additional protection is needed. If however, a safety source (see page 85) is not used then it is an Extra-Low Voltage system other than SELV, the exposed conductive parts of the circuit must be connected to the primary circuit protective conductor (471 - 14 - 03 to 05).

**Figure C5** **Segregation of Circuits – Categories 1, 2 and 3**

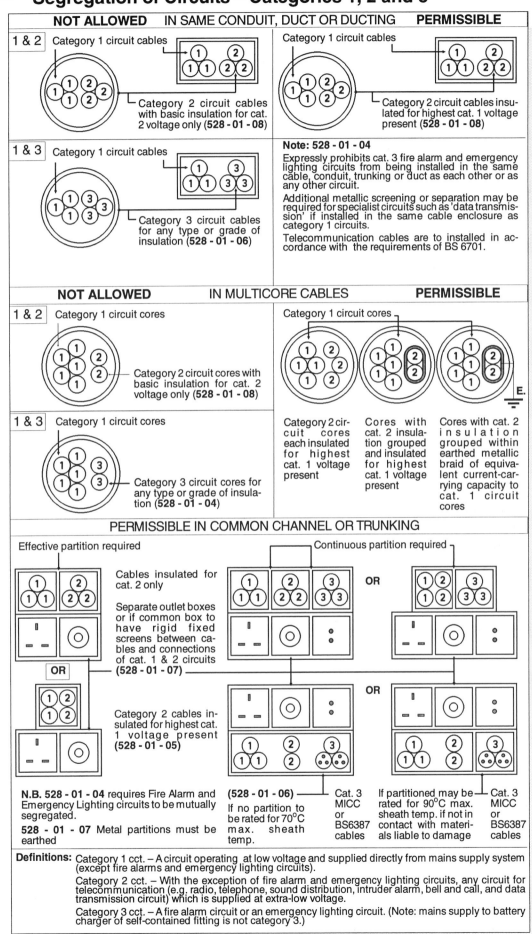

**NOT ALLOWED** IN SAME CONDUIT, DUCT OR DUCTING **PERMISSIBLE**

**1 & 2** Category 1 circuit cables

Category 2 circuit cables with basic insulation for cat. 2 voltage only (**528 - 01 - 08**)

Category 1 circuit cables

Category 2 circuit cables insulated for highest cat. 1 voltage present (**528 - 01 - 08**)

**1 & 3** Category 1 circuit cables

Category 3 circuit cables for any type or grade of insulation (**528 - 01 - 06**)

**Note: 528 - 01 - 04**
Expressly prohibits cat. 3 fire alarm and emergency lighting circuits from being installed in the same cable, conduit, trunking or duct as each other or as any other circuit.
Additional metallic screening or separation may be required for specialist circuits such as 'data transmission' if installed in the same cable enclosure as category 1 circuits.
Telecommunication cables are to installed in accordance with the requirements of BS 6701.

**NOT ALLOWED** IN MULTICORE CABLES **PERMISSIBLE**

**1 & 2** Category 1 circuit cores

Category 2 circuit cores with basic insulation for cat. 2 voltage only (**528 - 01 - 08**)

Category 1 circuit cores

E.

**1 & 3** Category 1 circuit cores

Category 3 circuit cores for any type or grade of insulation (**528 - 01 - 04**)

Category 2 circuit cores each insulated for highest cat. 1 voltage present

Cores with cat. 2 insulation grouped and insulated for highest cat. 1 voltage present

Cores with cat. 2 insulation grouped within earthed metallic braid of equivalent current-carrying capacity to cat. 1 circuit cores

PERMISSIBLE IN COMMON CHANNEL OR TRUNKING

Effective partition required

Continuous partition required

Cables insulated for cat. 2 only

Separate outlet boxes or if common box to have rigid fixed screens between cables and connections of cat. 1 & 2 circuits (**528 - 01 - 07**)

OR

OR

OR

Category 2 cables insulated for highest cat. 1 voltage present (**528 - 01 - 05**)

**N.B. 528 - 01 - 04** requires Fire Alarm and Emergency Lighting circuits to be mutually segregated.
**528 - 01 - 07** Metal partitions must be earthed

(**528 - 01 - 06**)
If no partition to be rated for 70°C max. sheath temp.

Cat. 3 MICC or BS6387 cables

If partitioned may be rated for 90°C max. sheath temp. if not in contact with materials liable to damage

Cat. 3 MICC or BS6387 cables

**Definitions:** Category 1 cct. – A circuit operating at low voltage and supplied directly from mains supply system (except fire alarms and emergency lighting circuits).
Category 2 cct. – With the exception of fire alarm and emergency lighting circuits, any circuit for telecommunication (e.g. radio, telephone, sound distribution, intruder alarm, bell and call, and data transmission circuit) which is supplied at extra-low voltage.
Category 3 cct. – A fire alarm circuit or an emergency lighting circuit. (Note: mains supply to battery charger of self-contained fitting is not category 3.)

# Protective Conductors and Earthing

**An Exposed Conductive Part** is a conductive part of equipment which can be touched, and which is not a live part, but which may become live under fault conditions. This definition refers to any metalwork which forms part of an electrical installation and applies to all metallic switchgear enclosures, machines, conduit, ducting, trunking, accessories, etc.

**An Extraneous Conductive Part** is a conductive part liable to introduce a potential, generally earth potential, and not forming part of the electrical installation. (See Note 1.)

**A Protective Conductor** is a conductor used for some measures of protection against electric shock, and intended for connecting together any of the following:

> exposed conductive parts;

> extraneous conductive parts;

> the main earthing terminal;

> earth electrode(s) or other means of earthing;

> the earthed point of the source, or an artificial neutral.

This is a general definition which covers the following functions:

**Bonding Conductor** is a protective conductor for ensuring equipotential bonding.

**Circuit Protective Conductor** (cpc) is a protective conductor connecting exposed conductive parts of equipment to the main earthing terminal.

**Earthing Conductor** is a protective conductor connecting the main earthing terminal of an installation to an earth electrode or other means of earthing.

Note 1: The definition of Extraneous Conductive Part was altered during the validity of the 15th Edition to clarify its intention. The previous reference to "liable to introduce a potential", and the words "generally earth potential" replaced "including earth potential". If a conductive part is unlikely to introduce either earth or any other potential, it can be touched without danger and, therefore, does not need to be bonded, even though it may be simultaneously accessible with an exposed or extraneous conductive part.

---

**See Part 2 of BS 7671: 1992 for the complete set of Definitions**

---

# TOPIC CHART 2 (Decision) Protective Conductors and Earthing

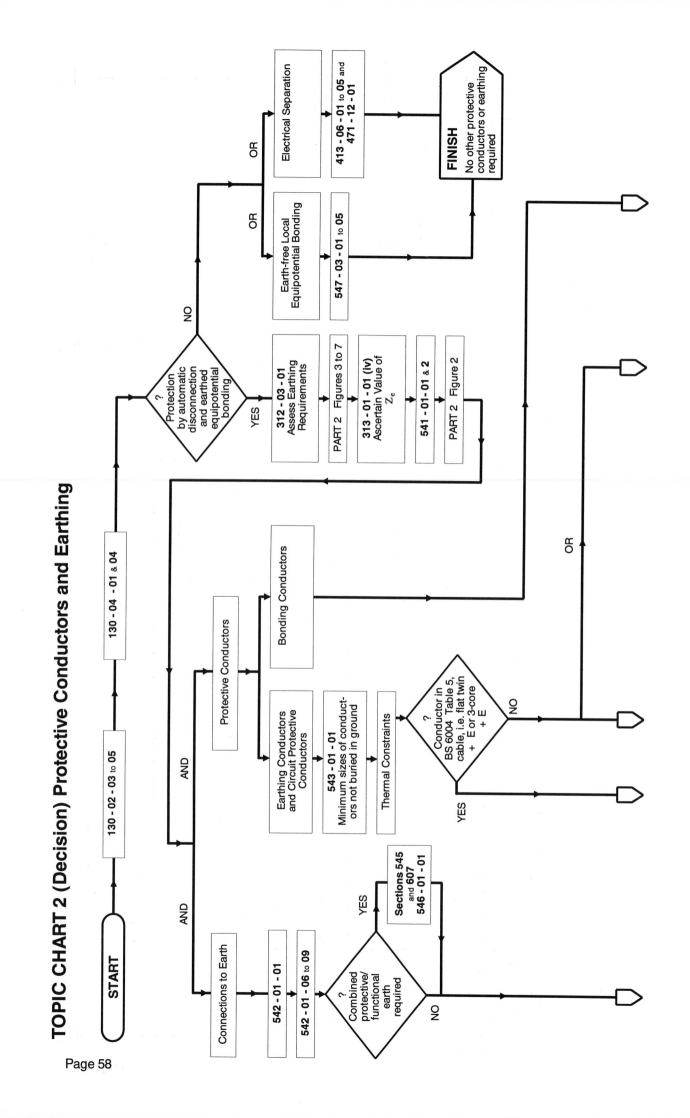

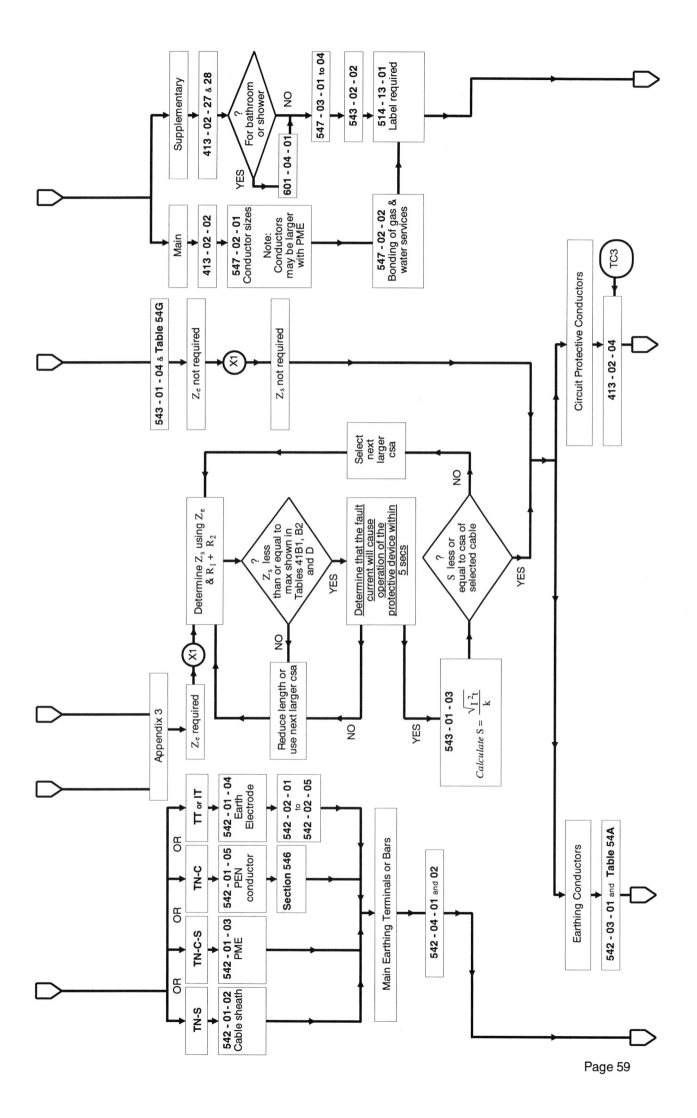

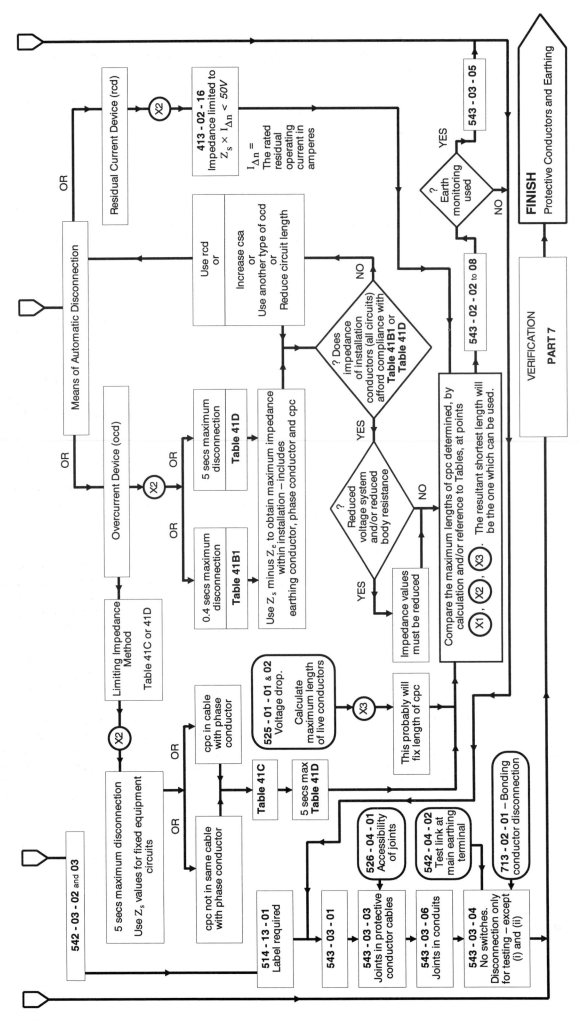

Note 1
continued:

For example, isolated door handles, coat hooks or shelf brackets are unlikely to introduce earth potential and are, therefore, not extraneous metal parts as defined, and do not need to be bonded.

On the other hand, metallic pipework for water, gas or oil is likely to be in contact with Earth, and must be deemed liable to introduce earth potential, so that appropriate main and perhaps supplementary bonding is required.

A generally acceptable test to establish whether an item is an extraneous conductive part is as follows:

Using a 500V d.c. insulation tester, measure the insulation resistance between the item and the main earthing terminal. If the resistance value is 0.25 megohm or greater, and inspection confirms that the resistance is unlikely to deteriorate, then the item can reasonably be considered not to be an extraneous conductive part.

The value of 0.25 megohm is based on the fact that 1mA flowing through the human body is not considered to be harmful, and this current will not be exceeded at the supply voltage with this resistance in circuit. Conversely if the insulation resistance test shows a negligible resistance, and this is confirmed by an ohm-meter as being 0.05 ohm or less, then the item is bonded and no further attention should be required.

BS 7671: 1992 regards supplementary equipotential bonding as being only required for ensuring the conditions for automatic disconnection (Regs. 413 - 02 - 08 to 14) are met, according to Regulation 413 - 02 - 15 and for meeting the requirements of Section 6.

Where supplementary equipotential bonding is required, it must connect together the exposed conductive parts of equipment in the circuit concerned. These connections include the earthing contacts of socket outlets and any extraneous conductive parts, see Regulations 547 - 03 - 01 to 05 which detail the requirements of the bonding conductor and its installation.

In locations containing a bath tub or shower basin, there are different requirements, and it is not Regulation 413 - 02 - 15 which must be satisfied, but Regulation 601 - 04 - 01 which demands supplementary bonding for all simultaneously accessible metal parts. This applies even if the room has no facility for the connection of fixed current-using equipment, since without bonding dangerous voltages could exist between extraneous conductive parts in the event of an earth fault elsewhere in the installation.

Regulations 547 - 03 - 01 to 03 give the sizing of supplementary bonding conductors. The minimum is $4mm^2$ if unprotected for connecting to exposed conductive parts. For joining extraneous conductive parts the minimum size when protected is $2.5mm^2$.

"If in doubt, bond it!" is sometimes thought to be the universal answer, but it may not be. This approach can sometimes actually introduce hazards. The best example is that of metal windows and patio doors. Under fault conditions a voltage can appear on the main earthing terminal, and hence on every metal part connected to it. This is not a hazard within the equipotential zone, but if the metal windows/doors are bonded they will also be subjected to this raised voltage and would therefore present a shock hazard to a person entering from outside, or to a window cleaner, in contact with earth.

Note 2:

A "main protective conductor" associated with, for example, a rising main feeding a sub-installation, may perform more than one specific function, and must therefore satisfy the requirements for each.

**There are four main factors which may affect the determination of the size of a protective conductor:-**

(a)     Physical strength and protection against mechanical damage and corrosion;

(b)     The capacity to carry earth fault current without thermal damage to itself and its surroundings;

(c)     Impedance limitations imposed by the 'protection against indirect contact' requirements of Regulations 413 - 02 - 08 and 09;

(d)     Voltage drop limits on associated live conductors where the protective conductor is a core within the same cable, or is within the same conduit, etc.

Factors (b), (c) and (d) may all result in limitations on the length of a protective conductor and it is necessary, therefore, to determine for each factor the related limitation on length.

The most onerous requirement of the three, i.e. the shortest of the three possible lengths, must be taken as the maximum permitted length of a protective conductor.

**All Protective Conductors** (Factor (a) above)

A main equipotential bonding conductor must have a cross-sectional area not less than 6mm$^2$ unless it is a TN-C-S system when the csa must not be less than 10mm$^2$.

If the supplementary bonding conductor is separate (i.e. if it is not part of cable, nor formed by conduit, ducting or trunking, nor within a wiring system enclosure) the cross-sectional area shall not in any event be less than:-

4 mm$^2$ if mechanical protection is not provided or

2.5mm$^2$ if mechanical protection is provided for connections between extraneous conductive parts.

Protective conductors of csa up to and including 10mm$^2$ must be of copper.

**Circuit Protective Conductors**

**A. Thermal Constraints** (Factor (b) above)

The cross-sectional area of protective conductors, other than equipotential bonding conductors, must be selected either by:-

(i) calculation in accordance with the equation in Regulation 543 - 01 - 03

$$S = \frac{\sqrt{I^2 t}}{k} \, mm^2$$

or (ii) application of Table 54G, in which case the cross-sectional area has a prescribed relationship to the associated phase conductor.

The application of (i) requires knowledge of the total earth fault loop impedance ($Z_s$) at the farthest point of a circuit. The nominal voltage divided by $Z_s$ then gives the value of earth fault current at the farthest point. By using the time/current characteristics given in Appendix 3 of BS 7671: 1992, the time (t) relating to the value of earth fault current can be obtained.

These values of earth fault current and time are then used in the equation with the relevant value of k taken from Tables 54B to F, to obtain the minimum permitted cross-sectional area for the protective conductor.

This value is unlikely to be exactly a standard size and it is necessary, therefore, to select the nearest larger standard protective conductor.

Note:  Reference should be made to BS 7430 which gives much more information on many aspects of earthing.

It must be emphasised that use of BS 6004 twin and cpc cables with reduced cpc requires application of the equation of Regulation 543 - 01 - 03. For values of $I^2t$ related to the protective device see Appendix 3 of BS 7671: 1992.

The application of (ii) is very simple, as the protective conductor is related to the associated phase conductor in accordance with Table 54G, i.e. where all conductors are of the same material:-

| phase conductor | protective conductor |
|---|---|
| $16mm^2$ or less | Same size |
| $25mm^2$ to $35mm^2$ | $16mm^2$ |
| over $35mm^2$ | Half size (to nearest higher standard csa) |

Where the protective conductor is not of the same material as the phase conductor, it must have the same conductance ratio to the phase conductor as that indicated above calculated by the formula in Table 54G.

This method automatically satisfies Factor (b) on page 62 and no knowledge of earth loop impedance or conductor length is required.

The method cannot be applied to BS 6004 twin and cpc cables with reduced cpc as, for conductors up to $16mm^2$, Table 54G requires the phase and protective conductor to be the same size.

If a circuit protective conductor is common to several circuits, its cross-sectional area is either to be calculated as described in sub-paragraph A(i) above, for the most unfavourable conditions of earth fault current and operating time in each of the associated circuits, or selected as described in sub-paragraph A(ii) according to the cross-sectional area of the largest associated phase conductor.  Such common protective conductors may be metal conduit or trunking containing more than one circuit, or a separate conductor in conduit or trunking.

**B.  Indirect Contact Protection Constraints** (Factor (c) on page 62)

Protective conductors must also comply with the 'protection against indirect contact' requirements of Section 413.

Whichever type of protection device is employed, it is necessary to comply with a limiting impedance for protective conductors which will often limit the length of a particular size of circuit protective conductor. Thus, it may be necessary to increase the protective conductor cross-sectional area to meet the requirements for shock protection.

The following alternative arrangements are available:

(i)     For circuits requiring disconnection in 0.4 seconds or less, for compliance with Regulation 413 - 02 - 09, the tabulated values of maximum $Z_s$ related to overcurrent protective devices in Table 41B1 and 41B2 apply. Although these values of impedance are identified for the disconnection of socket-outlet circuits in 0.4 seconds maximum, they must also be applied to circuits feeding fixed equipment in locations containing a bath tub or shower basin, which require disconnection in 0.4 seconds maximum (Regulation 601 - 04 - 02). In order to determine the earth fault loop impedance which can be tolerated within the installation (phase and protective conductors) it is necessary to deduct the external earth fault loop impedance ($Z_e$) from $Z_s$. Using the impedance per metre (for the actual cable being used, obtained from its manufacturer) for the phase and neutral conductors used in the installation the maximum length of cable can be calculated.

Fixed equipment supplied from a socket-outlet need not have a 4 second disconnection time if precautions are taken to prevent the use of the socket-outlet for supplying hand-held equipment.

**This application requires the use of the specific maximum value of $Z_e$.**

(ii) The application of Regulation 413 - 02 - 12 allows the disconnection time of a circuit supplying socket-outlets, portable equipment or hand-held Class I equipment to be increased to a maximum of 5 seconds. This entails compliance with the values of protective conductor impedance given in Table 41C according to the type and rating of the overcurrent protective device.

This offers advantages in the design of socket outlet circuits and should be used because:

(a) The degree of shock protection is unaffected if the fault is not of negligible impedance.

(b) The disconnection time can be increased from 0.4 seconds to 5 seconds.

(c) Larger values of external earth-loop impedance can be accepted.

Provided that the impedance of the protective conductor under fault conditions does not exceed the appropriate value in Table 41C, the shock voltage is limited to the internationally accepted "safe" value. Although the shock hazard is thus removed, the earth fault must not be allowed to persist indefinitely and it is also necessary, therefore, to comply with the relevant earth-loop impedance values in Table 41D to achieve disconnection within 5 seconds.

(iii) For fixed equipment circuits requiring disconnection in 5 seconds or less for compliance with Regulation 413 - 02 - 13, the tabulated values of maximum $Z_s$ related to overcurrent protective devices in Table 41D apply. The maximum length of cable can be calculated as above.

**This application requires the use of the specific maximum value of $Z_e$.**

(iv) For installations or circuits protected by a residual current device, the limits of $Z_s$ are determined by the application of Regulation 413 - 02 - 16.

The value of $Z_S$ associated with these devices is usually of a higher order than the value associated with overcurrent protective devices. There should be no problem in achieving satisfactory impedance within an installation. The minimum sizes of protective conductors are likely to be determined by the relevant BS for the cable or by the overriding minimum requirements relating to Factor (a) set out on page 62.

Note: Where conditions of reduced body resistance exist, or in the presence of livestock, it will be necessary to reduce the impedance values. It is recommended that maximum impedance values should be reduced to not more than 25 per cent of those shown in Regulation 413 - 02 - 10 and 11, or other methods of protection should be employed.

**C. Voltage Drop Constraints** (Factor (d) on page 62)

Where a protective conductor has a fixed ratio of cross-sectional area to its associated phase conductor throughout its length, and where the two conductors run together throughout the installation (e.g. as in a BS 6004 twin and cpc cable, or in a pvc conduit installation, etc.) then the volt-drop limitation on the length of the phase conductor of a final circuit protected by an overcurrent device rated at 100A or less also limits the length of the associated protective conductor.

Where the 4.0% voltage drop applies, it is likely that the consequent limitation on length will be more onerous on protective conductors than will be the limitations on length imposed by the thermal constraints of Regulation 543 - 02 - 04 or the shock protection constraints of Regulations 413 - 02 - 08 or 09.

It is recommended, therefore, that voltage drop limitations should be determined before the other two possible limits on a protective conductor.

Page 64

**Circuit Protective Conductors**

For domestic and similar installations, Regulation 413 - 02 - 12 is advantageous and, if it is possible to use multicore cables having a protective conductor with the same csa as the live conductors, then the adiabatic equation can be avoided by applying Regulation 543 - 01 - 04 and Table 54G.

Where protection against electric shock is provided by an overcurrent device the circuit protective conductor shall be incorporated in the same wiring system as the live conductors or in the immediate proximity of them (Regulation 544 - 01 - 01).

**Earthing Conductors**

In addition to meeting whichever of the foregoing requirements are relevant, earthing conductors are subject to additional requirements when buried in soil. Depending on whether or not an earthing conductor is protected against mechanical damage and/or protected against corrosion, it must satisfy Regulation 542 - 03 - 01 and the related Table 54A. Aluminium and copper clad aluminium must be protected against corrosion when used for earthing conductors with buried connections to earth electrodes. This is another requirement relating to Factor (a) on page 62.

**Bonding Conductors**

Requirements for sizing supplementary and main bonding conductors are contained in Section 547.

Whereas protective conductors are selected according to the level of earth fault current which they may carry, with equipotential bonding conductors it is virtually impossible to know exactly what earth fault current will flow.

Bonding conductors are selected, therefore, on the basis that the major proportion of an earth fault current will flow through the other protective conductors specifically intended to carry that current.

Main equipotential bonding conductors are required to have a cross-sectional area not less than half that of the related earthing conductor, subject to a minimum of 6mm$^2$ copper and a maximum of 25mm$^2$ copper, or equivalent conductivity.

Extraneous conductive parts are permitted to be used as protective conductors subject to Regulation 543 - 02 - 06 and all the requirements of Sub-section 542 - 02.

**Important note:**
Where a Protective Multiple Earthing (PME) service is provided by the Supplier, the main bonding connections must satisfy the requirements stated in the Electricity Supply Regulations 1988, as amended.

The cross-sectional area of the main bonding connections is to be related to the cross-sectional area of the supply neutral conductor (Regulation 547 - 02, Table 54H) but local public electricity supply network conditions may require a larger csa.

| Copper equivalent csa of supply neutral | Minimum copper equivalent csa of main bonding conductor |
|---|---|
| 35mm$^2$ or less | 10mm$^2$ |
| over 35mm$^2$ but not more than 50mm$^2$ | 16mm$^2$ |
| over 50mm$^2$ but not more than 95mm$^2$ | 25mm$^2$ |
| over 95mm$^2$ but not more than 150mm$^2$ | 35mm$^2$ |
| over 150mm$^2$ | 50mm$^2$ |

The cross-sectional area of a supplementary bonding conductor is determined by a reference to its function and the associated circuit protective conductor as shown below. In each case, the minimum permitted cross-sectional area is 2.5mm$^2$ where protected against mechanical damage, or 4mm$^2$ where not so protected.

| Supplementary bonding conductor connecting | Cross-sectional area not less than |
|---|---|
| Two exposed conductive parts (in bathrooms only) | Csa of smaller conductor connected to exposed conductive parts |
| Exposed and extraneous conductive parts | Half csa of protective conductor connected to exposed conductive part |
| Two extraneous conductive parts | 2.5mm$^2$ protected; 4mm$^2$ not protected against mechanical damage |
| Two extraneous conductive parts, one of which is bonded to an exposed conductive part | Half csa of protective conductor connected to exposed conductive part |

Regulation 547 - 03 - 05 permits the use of the circuit protective conductor within a flexible cord to provide the supplementary bonding connection.

**Earthing Arrangements**

Every installation must be provided with a main earthing terminal or bar for connecting;

> (i) the circuit protective conductors,

> (ii) the main bonding conductors, and

> (iii) any other protective conductors

to the earthing conductor of the installation.

Provision must be made for a means of disconnecting the main earthing conductor from the means of earthing. This can often be conveniently combined with the main earthing bar. This provision must:

> (i) be readily accessible,

> (ii) be electrically reliable,

> (iii) be mechanically sound, and

> (iv) require the use of a tool for disconnection.

Attention is drawn to the need to disconnect the main equipotential bonding conductors for measurements of $Z_e$. See also page 90 Isolation and Switching.

The earthing conductor of the installation must be connected to the relevant means of earthing of the system, e.g. TN-S, TN-C-S, TT.

Note: Only the Supplier may make the connection to the PME earthing terminal.

**Resistance of Conductors at Normal Operating Temperature**

Regulation 413 - 02 - 05 requires that account should be taken of the increase of temperature and resistance of circuit conductors as a result of overcurrent and goes on to state that if the protective device complies with Appendix 3 of BS 7671 and the loop impedance of the circuit is complying with Regulations 413 - 02 - 10, 413 - 02 - 11 and 413 - 02 - 14, the circuit is deemed to comply with the requirements of this Regulation.

However, tables 41B1, 41B2, and 41D have a note which requires the values of $Z_s$ should not be exceeded when the conductors are at their normal operating temperature.

This temperature is impossible to predict and it must be assumed that the normal operating temperature will not exceed the maximum allowable operating temperature of the particular type of cable.

Thus the value of $Z_s$ given in tables 41B1, 41B2 and 41D must be assumed to apply when the cable is operating at its maximum permissible operating temperature.

When tests are carried out to ensure compliance with this requirement the cables will not be at their operating temperature but will be at the ambient temperature. A temperature co-efficient must therefore be applied to the measured $Z_s$ before comparing it with the value given in the table.

The factor is calculated as follows:

$$1 + \left[ \left( \begin{array}{l} \text{Maximum temperature} \\ \text{that a cable can} \\ \text{operate at} \end{array} - \begin{array}{l} \text{Ambient temperature used} \\ \text{for design or at which} \\ \text{tests are carried out} \end{array} \right) \times 0.004 \right]$$

Thus for design based on or testing carried out at 20° C the factors for three of the most used cables are:

| | | |
|---|---|---|
| 70° C PVC | $1 + [(70° - 20°) \times 0.004]$ | $= 1.2$ |
| 85° C RUBBER | $1 + [(85° - 20°) \times 0.004]$ | $= 1.26$ |
| 90° C THERMOSETTING | $1 + [(90° - 20°) \times 0.004]$ | $= 1.28$ |

This used the simplified formula given in BS 6360 that the resistance–temperature coefficient is 0.004 for °C at 20° C for copper and aluminium conductors. Manufacturers will provide more accurate factors based on their own products.

The following example uses Manufacturers' typical values:

A circuit supplying socket outlets is wired with 2.5mm$^2$ 2-core pvc-insulated and sheathed cable incorporating a 1.5mm$^2$ protective conductor. The circuit is protected by a 20A fuse to BS88 Pt 2. The external earth loop impenance, $Z_e$, is 0.8 ohm

Determine the maximum length of run affording compliance with Regulation 413 - 02 - 08 and 09 – as follows:

### (a) Table 41B1 (a)

From table 41B1 (a), the maximum earth loop impedance, $Z_s$, associated with a 20A BS 88 fuse is 1.85 ohm.

As $Z_s = Z_e + (R_1 + R_2)$, where $(R_1 + R_2)$ is the resistance of the phase and protective conductors in series, the maximum permitted value of $(R_1 + R_2)$ is $1.85 - 0.8 = 1.05$ ohm.

A typical value of $(R_1 + R_2)$/metre is 19.51 milliohm/metre.

The typical multiplier for pvc insulation is 1.2.

Therefore the value of $(R_1 + R_2)$/metre under fault conditions is $19.51 \times 1.2 = 23.41$ milliohm/metre.

The maximum length of run affording

compliance with Table 41B1(a) is thus: $\dfrac{1.05}{0.02341} = 45$ metres

### Resistance of Conductors under Fault Conditions

Where Regulation 434 - 03 - 03 applies i.e. conductors are protected against fault current only or it is required to calculate the size of protective conductors in accordance with Regulation 543 - 01 - 03 it will be necessary to check for thermal constraints under fault conditions by applying the adiabatic equation as follows:

$$t = \frac{k^2 S^2}{I^2}$$

where $t$ = duration of fault in seconds
$I$ = fault current
$S$ = csa of conductor
$k$ = factor taking account of the heat capacity of the conductor

### Example:

A 63 amp BS88 Pt 2 fuse provides overload protection for 16mm$^2$ copper PVC insulated conductors fed from busbars which are protected by 200 amp BS88 Pt 2 fuses. The 16mm$^2$ tails between the 200 amp and 63 amp fuses are 4.5 m long. The prospective fault current at the 63 amp fuse is 1kA. Establish whether the 200 amp fuses will provide fault current protection for the 16mm$^2$ tails.

$$t = \frac{k^2 S^2}{I^2}$$

k = 115 from Table 43A BS7671
S = 16mm$^2$
I = 1000 amps

$$t = \frac{115^2 \times 16^2}{1000^2}$$

t = 3.4 seconds

From Figure 3A (Appendix 3) BS7671 at 1000 amps a 200 amp BS88 fuse will disconnect in 7.5 seconds. Since the maximum permitted time (t) is only 3.4 secs it can be seen that the 200 amp fuse does *not* protect the tails against fault current.

A further calculation with a larger size of cable (25mm$^2$) would therefore be required.

$$t = \frac{115^2 \times 25^2}{1000^2}$$

t = 8.2 seconds

Since the maximum duration time (t) has risen to 8.2 secs and the disconnection time is 7.5 secs 25mm$^2$ cable *is* protected against fault current.

*Note*: From Regulation 473 - 02 - 02 if the length of tails is limited to 3 metres and erected in such a manner as to minimise the risk of fire or danger to persons then the above calculation need not be applied.

**Earthing Arrangements for Combined Protective and Functional Purposes**

Functional earthing is defined as "earthing necessary for the proper functioning of equipment", and Regulation 331 - 01 - 01 requires that an assessment be made of possible harmful effects of this requirement. Regulation 542 - 01 - 06 allows that functional earthing may be either combined with, or separated from, protective earthing according to the requirements of the installation.

Section 607 deals with these matters in particular as they affect data processing equipment to BS 6204. There are two main divisions of earthing requirement. Regulation 545 - 01 - 01 requires that if there is a need for a "clean", low noise, earth, then this must be discussed and agreed with the equipment manufacturers, and it may be that combined protective and functional earthing is not possible.

However, if there is no such consideration then Regulations 607 - 02 - 01 to 07 (inclusive) detail the methods to be adopted, depending on whether the equipment has a normal leakage current of:

(a) less than 3.5mA, when no special requirements apply

(b) more than 3.5mA and less than 10mA when the connection of fixed wiring shall be either permanent or via BS EN 60309-2 plugs and sockets

(c) more than 10mA, when one or more of six methods of high integrity protective connections must be adopted, and the connection to the supply shall preferably be permanent although BS EN 60309-2 plugs and sockets are acceptable. Alternatively a monitoring system to BS 4444 (which disconnects the supply in accordance with regulation 413 - 02 - 01 in the event of a discontinuity in the protective conductor) may be used.

Regulations 607 - 02 - 03 and 607 - 03 - 01 give the requirements when rcd protection is provided, and it is pointed out that it may be necessary to supply such equipment from an isolating Transformer. This is a relatively specialised field, and a thorough study of the relevant Regulations is recommended.

**Figure C6(a)**

# General Requirements for Protective Conductors
### (for sizing see Figures C6(b) and D9)

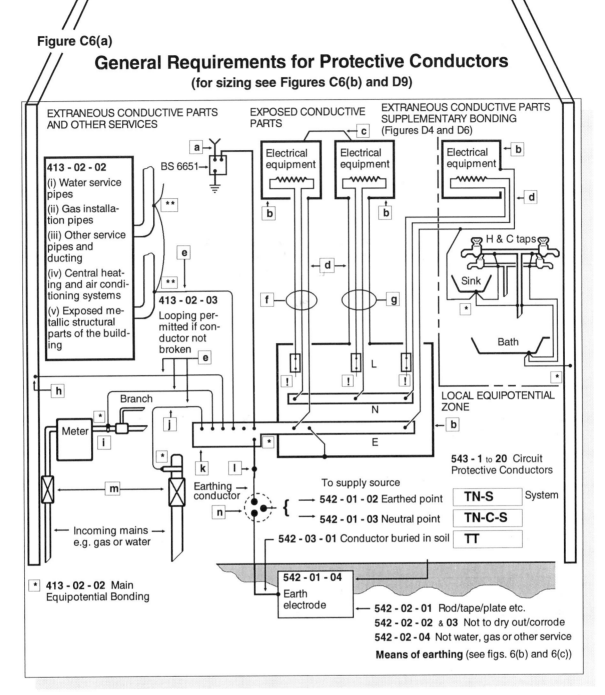

a. Lightning conductor
b. Exposed conductive parts
c. Supplementary bond
d. Protective conductor which can also be the supplementary bonding conductor if a flexible cord is used to feed the fixed equipment (547 - 03 - 05)
e. Main equipotential bonding (413 - 02 - 02)
f. Composite cable (543 - 02 - 02)
g. Conduit duct/trunking (543 - 02 - 02)
h. Exposed metallic parts of building structure
i. Maximum distance 600mm from service meter for equipotential bonding connection (547 - 02 - 02)
j. Not Al or CuAl if subject to corrosion
k. Main earth terminal (542 - 04 - 01)
l. Accessible means of disconnection by tool only (542 - 04 - 02)
m. Insulating section (if any) all bonding to be on consumer's side (547 - 02 - 02)
n. Connection of earthing conductor to means of earthing protected against corrosion (542 - 03 - 03)

\* Accessible connection
  543 - 03 - 03 Accessible connection and
    514 - 13 - 01 Warning notice
    SAFETY ELECTRICAL CONNECTION
        DO NOT REMOVE
\*\* Accessible and unbroken connection
! Protective devices. Characteristics and earth fault loop impedance ($Z_s$) of circuits to be coordinated so that:

– socket-outlet circuits are disconnected in 0.4 seconds or less
– fixed equipment circuits are disconnected in 5.0 seconds or less
(Regs. 413 - 02 - 04 to 09 and Tables 41B1 to 41D)

and for rcds $Z_s$ x tripping current should be less than 50 volts (413 - 02 - 16)

Note: BS 951 compliance necessary for all connections clamped to pipes

# Sizing: Protective Conductors; Earthing Conductors; Main Equipotential Bonding Conductors; Circuit Protective Conductors

**543 - 01 - 01** and **02**

Calculation or selection of protective conductor cross-sectional area – equipotential bonds excluded

**543 - 01 - 03**

To be determined as:-

S(csa)mm² of protective conductor $= \dfrac{\sqrt{I^2 t}}{k}$

    t = disconnection time

    k = Material factor from Reg. 543 - 01 - 03 of the Regulations

    I = Fault current

    **OR**

Size to be related to that of phase conductor (S mm²) – see table 54G

Where S is equal or less than 16mm² protective conductor to be equal to S

    S is greater than 16mm² and less or equal to 35mm² protective conductor to be 16mm²

    S is greater than 35mm² protective conductor to be half the phase conductor csa

If protective and phase conductor are not of same material then conductance of protective conductor to be not less than that conductance it would have if it were of the same material and selected as stated above.

**NOTE:**

BS 6651 (Protection of Structures against Lightning) requires that a main equipotential bond be extended to the lightning conductor, via as short a route as possible to the conductor side of the test clamp. BS 6651 also requires bonding to liftshaft metal, e.g. guides, handrails, ladders, etc. at top and bottom of the shaft and additionally at intervals of not more than 20m which is more than 'extraneous metal' would require for 16th Edition compliance.

See Tables 41B1, 41B2, 41C and 41D for calculated values of maximum earth loop impedance relating to BS 6004 cables which take into consideration the maximum value of t, i.e. 0.4 or 5 seconds.

**PROTECTIVE CONDUCTORS MAY COMPRISE THE FOLLOWING (543 - 02 - 02)**

Single core cable

Conductor in a cable

Conductor in a common enclosure with insulated live conductors

Fixed bare or insulated conductor

Metal sheath, etc. of a cable

Metal conduit or enclosure

An extraneous conductive part

**MINIMUM SIZE OF SEPARATE PROTECTIVE CONDUCTOR, NOT CONDUIT, ETC.**

**543 - 01 - 01**

Minimum cross-sectional area not less than 2.5mm² copper equivalent or 4mm² if not mechanically protected – see also 543 - 03 - 01

**or if:-**

**BURIED IN GROUND**

**542 - 03 - 01**, Table 54A and BS 7430

This shall apply as if the protective conductor were an earthing conductor

If protected against corrosion but not mechanical damage, the minimum cross-sectional area is 16mm²

N.B.

If related to a PME system, the special requirements must be complied with (check with the Supplier)

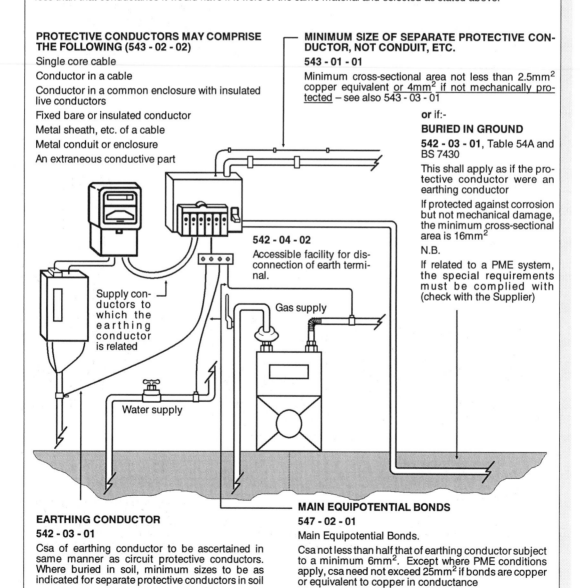

**542 - 04 - 02**

Accessible facility for disconnection of earth terminal.

Supply conductors to which the earthing conductor is related

Gas supply

Water supply

**EARTHING CONDUCTOR**

**542 - 03 - 01**

Csa of earthing conductor to be ascertained in same manner as circuit protective conductors. Where buried in soil, minimum sizes to be as indicated for separate protective conductors in soil

**MAIN EQUIPOTENTIAL BONDS**

**547 - 02 - 01**

Main Equipotential Bonds.

Csa not less than half that of earthing conductor subject to a minimum 6mm². Except where PME conditions apply, csa need not exceed 25mm² if bonds are copper or equivalent to copper in conductance

**Figure C6(c)**

# The Seven Recognised Types of Earth Electrode (542 - 02 - 01)

**ALUMINIUM**

Aluminium and copper-clad not to be used for final connection to earth electrodes <u>where corrosion may occur</u>

**CONNECTION**

**542 - 03 - 03**

Connection to earth electrode to be labelled, soundly made, protected against corrosion and electrically and mechanically satisfactory (See 514 - 09 - 01 for identification)

**MATERIALS & CONSTRUCTION**

**542 - 02 - 03**

Materials used and the construction must be such as to withstand damage

**CORROSION**

**542 - 02 - 03**

Design of earthing arrangements to take account of possible increase in earth resistance due to corrosion

**DRYING / FREEZING**

**542 - 02 - 02**

Type and embedded depth must be such that soil drying or freezing will not increase resistance unacceptably

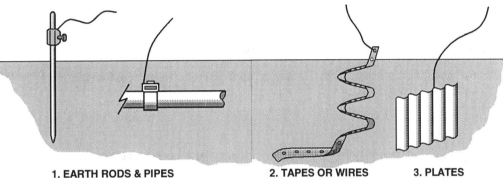

**1. EARTH RODS & PIPES**

**542 - 02 - 04**

Gas and water services not to be used as protective earth electrodes

**2. TAPES OR WIRES**

**3. PLATES**

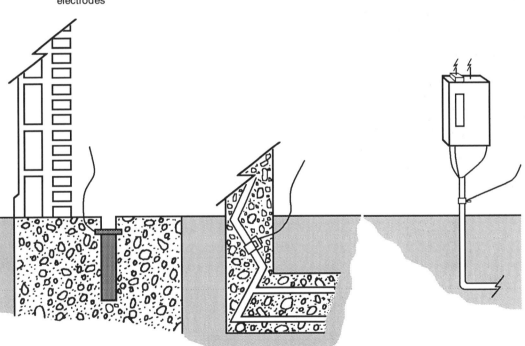

**4. UNDERGROUND STRUCTURAL METALWORK EMBEDDED IN FOUNDATIONS**

See BS 7430 for dimensional details of earth electrodes

**5. WELDED METAL REINFORCEMENT OF CONCRETE (NOT PRE-STRESSED)**

**7. OTHER SUITABLE UNDERGROUND METALWORK** (not illustrated)

**6. LEAD OR METALLIC COVERINGS OF CABLES**

**542 - 03 - 01**

lead or metallic covering must be in effective contact with the ground. Consent of owner to be obtained. Arrangements for installation owner to be notified of effective changes which may be proposed to be made to the cables.

# Topic Chart 3

# Protection against Electric Shock

**Direct Contact.** Contact of persons or livestock with live parts.

**Indirect Contact.** Contact of persons or livestock with exposed-conductive-parts which have become live under fault conditions.

**Live Part.** A conductor or conductive part intended to be energised in normal use, including a neutral conductor but, by convention, not a PEN conductor.

**SELV.** An Extra-low voltage system which is electrically separated from earth and from other systems in such a way that a single fault cannot give rise to the risk of electric shock.

**PELV (protective extra-low voltage).** An extra-low voltage system which is not electrically separated from earth, but which otherwise satisfies all the requirements for SELV.

---

**See Part 2 of BS 7671: 1992 for the complete set of Definitions**

# TOPIC CHART 3 (Information) Protection against Electric Shock

START

471 - 13 - 01      Exempts skilled persons, subject to Electricity at Work Regulations 1989

130 - 04 - 03

410 - 01

**OR**

**AND**

## Protection against Both Direct and Indirect Contact

411 - 01 - 01

### SELV

411 - 02 - 01
Protection

411 - 02 - 02 to 04
Sources

411 - 02 - 05 to 11
Circuit arrangements

471 - 02 - 01

Section 6

### Extra-Low Voltage systems other than SELV

411 - 03 - 01
Not to be used alone

471 - 14 - 01 to 06

### Limit energy discharge

471 - 03 - 01

411 - 04 - 01

## Protection against Indirect Contact

### Insulation of live parts

471 - 04 - 01

412 - 02 - 01

713 - 05 - 01 & 02
Requirements to test site applied insulation

### Barriers or enclosures

471 - 05 - 01

412 - 03 - 01
Requires IP2X or better

471 - 05 - 02
Exemption from IP2X lampholders, etc.

412 - 03 - 02 to 04

### Obstacles

471 - 06 - 01

412 - 04 - 01 & 02

### Placing out of reach

471 - 07 - 01

412 - 05 - 01

IEE Guidance Notes No 5 Protection Against Electric Shock

412 - 05 - 02 to 04

## Protection against Direct Contact

412 - 01 - 01

### Residual current devices

412 - 06 - 01 & 02
Supplementary protection

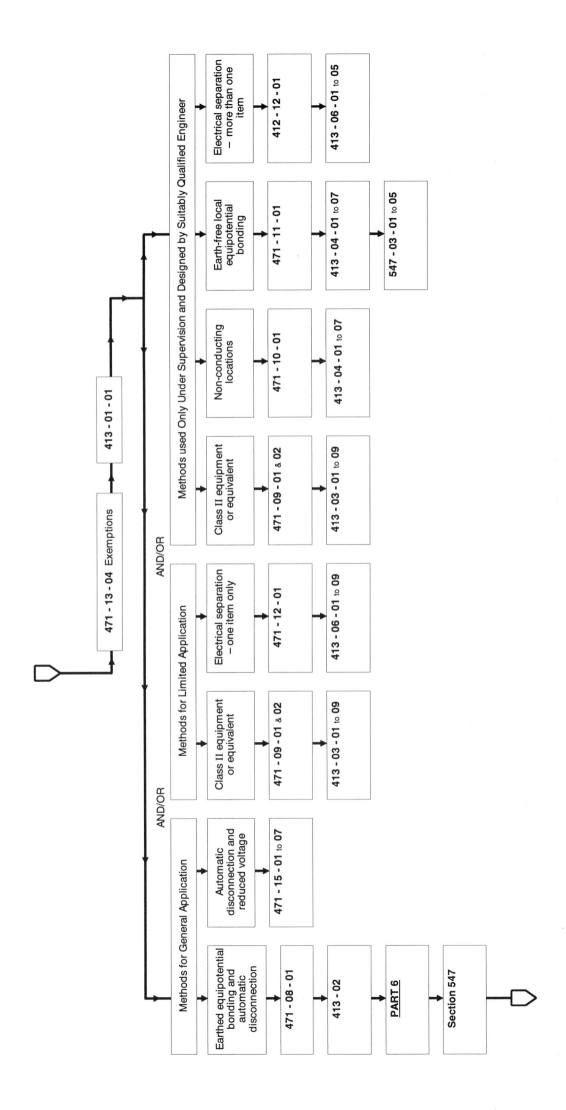

471 - 13 - 04  Exemptions

413 - 01 - 01

**Methods for General Application**

Earthed equipotential bonding and automatic disconnection

Automatic disconnection and reduced voltage

471 - 08 - 01

471 - 15 - 01 to 07

413 - 02

PART 6

Section 547

AND/OR

**Methods for Limited Application**

Class II equipment or equivalent

Electrical separation – one item only

471 - 09 - 01 & 02

471 - 12 - 01

413 - 03 - 01 to 09

413 - 06 - 01 to 09

AND/OR

**Methods used Only Under Supervision and Designed by Suitably Qualified Engineer**

Class II equipment or equivalent

Non-conducting locations

Earth-free local equipotential bonding

Electrical separation – more than one item

471 - 09 - 01 & 02

471 - 10 - 01

471 - 11 - 01

412 - 12 - 01

413 - 03 - 01 to 09

413 - 04 - 01 to 07

413 - 04 - 01 to 07

413 - 06 - 01 to 05

547 - 03 - 01 to 05

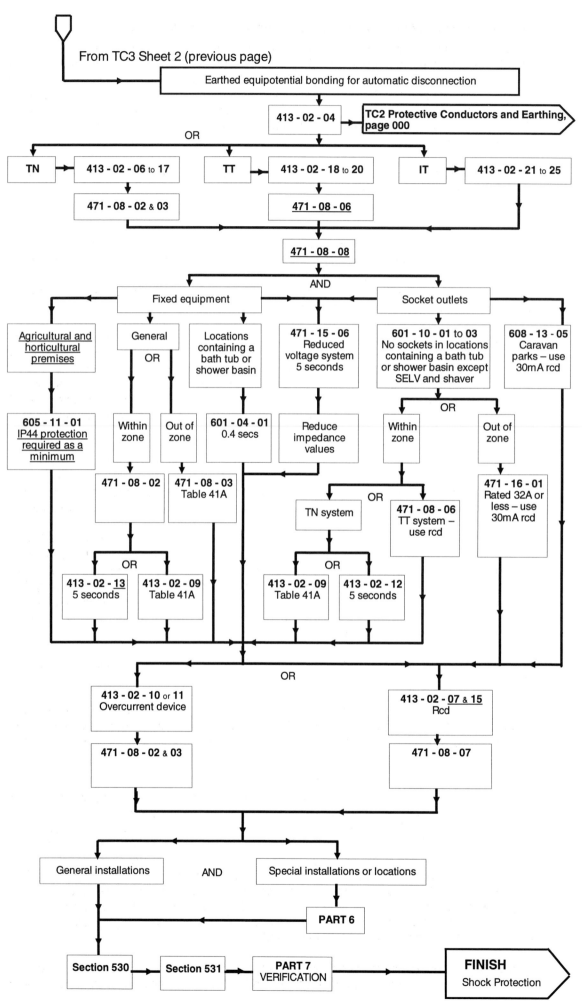

From TC3 Sheet 2 (previous page)

Earthed equipotential bonding for automatic disconnection

413 - 02 - 04 → **TC2 Protective Conductors and Earthing, page 000**

OR

**TN** → **413 - 02 - 06** to **17**
**471 - 08 - 02** & **03**

**TT** → **413 - 02 - 18** to **20**
**471 - 08 - 06**

**IT** → **413 - 02 - 21** to **25**

**471 - 08 - 08**

AND

Fixed equipment

Socket outlets

Agricultural and horticultural premises

General
OR

Locations containing a bath tub or shower basin

**471 - 15 - 06** Reduced voltage system 5 seconds

**601 - 10 - 01** to **03** No sockets in locations containing a bath tub or shower basin except SELV and shaver

**608 - 13 - 05** Caravan parks – use 30mA rcd

**605 - 11 - 01** IP44 protection required as a minimum

Within zone

Out of zone

**601 - 04 - 01** 0.4 secs

Reduce impedance values

OR

Within zone

Out of zone

**471 - 08 - 02**

**471 - 08 - 03** Table 41A

TN system

OR

**471 - 08 - 06** TT system – use rcd

**471 - 16 - 01** Rated 32A or less – use 30mA rcd

OR

**413 - 02 - 13** 5 seconds

**413 - 02 - 09** Table 41A

OR

**413 - 02 - 09** Table 41A

**413 - 02 - 12** 5 seconds

OR

**413 - 02 - 10** or **11** Overcurrent device

**413 - 02 - 07** & **15** Rcd

**471 - 08 - 02** & **03**

**471 - 08 - 07**

General installations

AND

Special installations or locations

**PART 6**

**Section 530** → **Section 531** → **PART 7 VERIFICATION** → **FINISH** Shock Protection

## Combined Protection against Both Direct and Indirect Contact

The scope available to the designer for taking advantage of combined protection is extremely limited, being confined to circuits operating at safety extra-low voltage or a circuit for which the supply is derived from a source having special characteristics. Functional extra-low voltage must not be used alone (411 - 01 - 01) but may be used in combination with other protective measures. These circuits are intended for use in areas of high hazard, e.g. where persons are in damp situations or in close contact with earthed surroundings, such as whilst working inside boilers and pressure vessels (see also Section 606, Restrictive Conductive Locations). It is likely to be necessary to provide combined protection against direct and indirect contact where lighting is provided by handlamps. Such lamps in garages should be shielded against breakage and possible subsequent ignition of fuel.

## SELV

The characteristics of a safety extra-low voltage system are that the voltage shall not exceed 50V rms a.c. or 120V ripple-free d.c., precautions having been taken to ensure that the voltage cannot rise above that limit, and that no conductors in the circuit may be connected either deliberately or accidentally with earth, and guarding against accidental short-circuit of high current/low voltage conductors will be necessary to avoid burns or other injuries.

If the voltage of the SELV system is maintained at 25V a.c. (60V d.c.) or less, it is not necessary to provide protection against direct contact for the live parts, i.e. bare conductors are permitted (insulation or physical separation will be required, of course, for functional reasons). Above 25V a.c. (60V d.c.) and up to 50V a.c. (120 d.c.) it is necessary to provide insulation or IP2X barriers or enclosures. Protection against overcurrent is however still required, and guarding against accidental short-circuiting of high current low voltage will be necessary to avoid burns and other injuries.

All of the above voltage limits must be reconsidered and reduced if the location of the installation is such that presence of dampness, water or excessive temperature is to be expected, as in these conditions the resistance of the human body is significantly lowered and the likelihood of dangerous shock increased. These special installations or locations are specified in Part 6.

To ensure that SELV conductors are not earthed accidentally, careful design and installation techniques are required to ensure that separation is effective and maintained. Physical separation is the preferred method, and this implies a totally separate installation with no exposed conductive parts connected to earth, to other conductors or metalwork. Thus, it is unlikely that a metal conduit or trunking installation would satisfy this requirement because the possibility of a fixing screw coming into contact with earthed metal cannot be discounted, and neither must the problem of a steel-framed building be ignored.

SELV installations should, therefore, be carried out in insulated conduit or trunking or non-metallic sheathed cables without protective conductors.

SELV plugs and socket-outlets must be of an exclusive pattern, throughout the premises involved. The plugs must not be able to enter (even partially) any other socket-outlet, and the socket-outlets must not accept any other plug or contain provision for connection of a protective conductor.

## Functional Extra-low voltage systems

In some installations, extra-low voltage may be selected for operational reasons, but it may not be possible to satisfy all the requirements for SELV circuits. These systems are termed PELV (471 - 14 - 02) when all the requirements of SELV are met except that the circuits are earthed at one point and 'Functional extra low voltage' where other exceptions to SELV are required, and may, for example, occur where metallic conduit is employed for protection against mechanical damage where Class I equipment is used. In addition, certain components of an extra-low voltage circuit, such as relays, may not have the degree of separation required between contacts carrying extra-low voltage and those contacts or parts connected to a higher voltage circuit; the degree of separation necessary being equivalent to that of a safety isolating transformer to BS 3535. (See the requirements of 471 - 14 - 01 to 06.)

If the system proposed complies with all the requirements for safety extra-low voltage, except that live or exposed conductive parts are connected to earth or to the protective conductors of other systems (e.g. one conductor connected to earth or a metallic conduit system) then no additional protection against indirect contact is required, but enclosures to at least IP2X (BS 4590) must be provided, or insulation of live parts must be subjected to an a.c. test voltage or 500V rms for one minute without failure or deterioration.

If there are other features of the installation which do not comply with the safety extra-low voltage requirements, then further protection against both direct contact and indirect contact is required by insulation and/or barriers or enclosures.

In such cases, barriers or enclosures must be provided to prevent unintentional contact, and these are specified in Regulations 412 - 03 - 01 to 04.

Insulation must be suitable for the minimum test voltage of the primary circuit and where accessible, it must be reinforced during erecting to withstand a test voltage of 1500V rms for one minute. Exposed conductive parts must be connected to the protective conductor of the primary circuit. It follows that such a circuit, although operating at extra-low voltage, must comply with the requirements for a low voltage installation.

If socket-outlets are installed in extra-low voltage circuits other than SELV, they must be of an exclusive pattern different from those in SELV circuits within the same premises, and may contain provision for the connection of a protective conductor.

If any luminaire support coupler has a contact for a protective conductor, it must not be used on a SELV circuit.

**NOTE:** SELV circuits will only be acceptable under the Electricity At Work Regulations, 1989, provided that all portable equipment is of Class II or Class III insulated construction.

## Limitation of Discharge Energy

The Regulations recognize that protection against shock can be afforded if the source of energy is controlled or designed so that the current which can pass through the body of a person or livestock is lower than the shock current, i.e. that current likely to cause danger. There is no limit imposed on the open circuit voltage.

This method has limited application, the best known of which is an electric fence controller. It must be ensured that the source of energy complies with the relevant British Standard for such equipment. It should not be assumed, however, that since the energy discharge itself should not cause danger, there will be negligible risk – e.g. BS 2632 requires for electric fence controllers, a peak voltage between 1.5 and 10 kV and a peak current between 150 mA and 10 A. The time pulse is, however, limited to 0.1 seconds, and the current may only exceed 300 mA for 1.5 milliseconds.

Circuits employing limitation of discharge energy as a means of protection against shock must be separated from other circuits as required by Regulations 411 - 02 - 05(i), 05(ii) and 11 for safety-low voltage circuits.

## Protection against Direct Contact

The measures which may be adopted to prevent direct contact are one or more of the following:

(i) Protection by insulation of live parts (Regulation 412 - 02 - 01)

(ii) Protection by barriers or enclosures (Regulations 412 - 03 - 01 to 04).

(iii) Protection by obstacles (Regulations 412 - 04 - 01 and 02)

(iv) Protection by placing out of reach (Regulations 412 - 05 - 01 to 04)

It is extremely unlikely that any complete installation encountered in ordinary circumstances will rely on only one of the above measures. It is probable that at least two, protection by insulation and by barriers or enclosures will be used most often, depending on the particular part of the installation being considered.

Protection by insulation of live parts includes insulated cables. Enclosures may be conduit, metal boxes, or insulated boxes where it is only possible to gain access by use of a key or tool such as a screwdriver.

In certain circumstances, e.g. the contacts within a distribution fuse-board, it may not be practical to insulate or enclose live parts. In these circumstances, barriers (e.g. resin bonded board) may be used to prevent unintentional contact; but by definition, will not, and in this example, must not, prevent deliberate action.

It is important to understand that where two methods are used together, one of the methods must meet the requirements fully. For example, if a fuse-board has insulated busbars and connections, the enclosure may be opened without the use of a key or tool. However, it is essential that all live parts within the enclosure be fully shrouded. In other words, the partial application of two different measures of protection cannot be combined to claim compliance with the degree of protection required.

It is important to note the requirements of 471 - 05 - 02. These qualify Regulation 412 - 03 - 01 and allow openings greater than those permitted IP2X, in order to permit items where those requirements are impracticable. The obvious example given is lampholders, which are currently exempted from the Electrical Equipment (Safety) Regulations.

An example of the use of obstacles to prevent direct contact is a handrail in front of an open-type switchboard, but the use of obstacles is applicable where only skilled or instructed persons, under direct supervision, have access (Regulation 471 - 06 - 01). However, certain conditions in Part 6 limit the use of this method and means not only objects in regular use but occasional use such as metal ladders must be taken into account.

Placing out of reach is a method of protection with somewhat restricted application to locations accessible only to skilled or instructed persons. Where bulky or long conducting objects may be handled, it is inadvisable to rely upon this method.

## Protection against Indirect Contact

Five basic protective measures for protection against indirect contact (see definition) are specified:

    (i) Earthed equipotential bonding and automatic disconnection of supply (Regulations 413 - 02 - 02 to 28 and 471 - 08 - 01 to 471 - 08 - 08.

    (ii) Use of Class II equipment or equivalent insulation (Regulation 413 - 03 - 01 to 09 and 471 - 09 - 01 to 471 - 09 - 04).

    (iii) Non-conducting location (Regulations 413 - 04 - 02 to 07 and 471 - 10 - 01).

    (iv) Earth-free local equipotential bonding (Regulations 413 - 05 - 01 to 04 and 471 - 11 - 01).

    (v) Electrical separation (Regulations 413 - 06 - 01 to 05 and 471 - 12 - 01).

The principles underlying the first method are those which have been established practice in the U.K. over many years, and earthed equipotential bonding and automatic disconnection of supply remains the preferred method of protection against indirect contact.

The other four deal with special installations likely to be under the control of trained and experienced persons only. These installations are unlikely to be encountered in normal practice, and they are dealt with later.

## Earthed Equipotential Bonding and Automatic Disconnection

The purpose of the bonding of all extraneous conductive parts is to create an equipotential zone within which all voltages between exposed conductive parts and extraneous conductive parts are minimized during earth fault conditions. See Notes 1 & 2 (pages 57 and 61) of this Handbook for an explanation of the necessity or otherwise to bond extraneous conductive parts.

The earthing arrangements for the installation, and the impedance of the protective conductors must be co-ordinated with the operating characteristics of the protective device for automatic disconnection of the supply, so that under earth fault conditions the prospective shock voltages within the equipotential zone are limited in magnitude and time duration. The achievement of this co-ordination by using, for example, the limiting values of impedances contained in Tables 41B, 41D Regulations 413 - 02 - 10 to 14 is dealt with in the notes accompanying Topic Chart 2 - "Protective Conductors and Earthing".

The designer and installer must ensure that the earth fault loop impedance is sufficiently low for the disconnection to occur within 0.4 seconds for final circuits supplying socket-outlets, and 5 seconds for final circuits supplying only fixed appliances. The reason for the difference is the assumption that socket outlets are normally used to supply portable and hand-held appliances, so that there is a greater risk of a person being in contact with the appliance at the time of the fault.

It should also be appreciated that the voltage difference between any part of the equipotential zone and "true" earth may be sufficient to cause a dangerous shock current.

It is important to note that the application of the limiting values of earth fault loop impedance contained in Regulations 413 - 02 - 10 and 11and limiting impedances in 413 - 02 - 12 are only applicable to circuits where all exposed conductive parts of the equipment concerned, and any extraneous conductive parts, are within the main equipotential zone.

Where an installation incorporates socket outlets rated at 32A or less, which may reasonably be expected to supply portable equipment for use outdoors, Regulation 471 - 16 - 01 requires a residual current device (30mA or less operating current and an operating time of not more than 40ms at a residual current of 150mA) in addition to the disconnection time of 0.4 seconds required by 471 - 08 - 02 and Table 41A. For supplies to fixed equipment, a disconnection time of 0.4 seconds or less is required, and this may be achieved by use of an overcurrent protective device or an rcd.

Examples of the application of these requirements are (i) in domestic situations where socket outlets are provided for garden tools, or socket outlets adjacent to openable windows from which garden tools etc. may be fed. Industrial situations where there are socket outlets for the connection of wire brushes, drills, etc., for external use; and (ii) in an industrial situation, such as an outside hoist where the control device may be used inside or outside the zone.

In all installations connected to TT systems, every socket-outlet circuit must be protected by a residual current device, although not necessarily individually.

Residual current devices of the specified operating current and time are only to be considered as additional protection against direct contact. At least one of the measures specified for protection against direct contact must also be used. Devices having the prescribed characteristics may provide additional protection in situations such as domestic gardens where flexible cords for mowers, hedge trimmers, etc., are liable to damage.

Reference should be made to BS 4293 for further information on residual current operated circuit breakers.

## Class II Equipment (BS 2754) or Equivalent Insulation

This method relies on the provision of double or reinforced insulation, to present a dangerous voltage appearing on exposed conductive parts through the failure of basic insulation. Class II (and for that matter Class I) is a term which applies to appliances.

Apart from Class II fixed appliances forming part of an installation, some accessories and other equipment may be of a construction equivalent to Class II. Where, for example, insulated plate-switches are installed, a circuit protective conductor must be provided since there is the possibility that one might later be replaced by a metal accessory.

Installations where it can be safely assumed that only Class II equipment will ever be installed or used will occur very rarely, and only in special circumstances where they are under effective supervision in normal use.

Class II equipment should not be earthed either deliberately or fortuitously because this could impair the protection against indirect contact afforded by the equipment specification.

Moreover, if it is desired to mount Class II equipment on earthed metalwork, e.g. a Class II luminaire mounted on a metal conduit box, it must be ensured that basic insulation is not the only protection between live parts of the installation and the exposed metalwork of the Class II equipment. If the mounting face of the Class II equipment is metallic, this requirement implies the provision of an insulating separator between the equipment and the metal box.

The explicit requirements of these Regulations for the provision and testing of insulating enclosures, particularly if they are site constructed, require special consideration and experience (see Inspection and Testing on page 150).

## Note on Classification of Means of Protection Against Electric Shock

Full information on the various classifications is given in BS 2754 to which reference should be made, but a summary of the methods is given below. This BS is based on I.E.C. Report No.536 and provides for four classifications which do not indicate the degree of protection, but only the method of achieving it.

Class 0, in which protection is provided only by basic insulation; no provision is made for the earthing of accessible conductive parts. Such an arrangement would only be acceptable in an earth-free situation and since this cannot be continuously guaranteed, Class 0 is **not** accepted in the United Kingdom, and there are statutory prohibitions against its use.

Class I relies on basic insulation, and the earthing of all accessible conductive parts. It follows that to comply with the BS, the flexible cord of Class I equipment must include a protective conductor, connecting the accessible conductive parts to the circuit protective conductor.

Class II equipment does not rely solely on basic insulation, but has double or reinforced insulation. No provision is made for the connection of a protective conductor to the accessible conductive parts, (if any).

Class III equipment relies for protection on the use of Safety Extra-Low Voltage (SELV) supplies, and no voltages higher than ELV may be generated within the equipment. (ELV is a.c. of not more than 50V, or d.c. not exceeding 120V and ripple-free. See the appropriate sections of this Handbook for further information.)

**N.B.**
Because of the Electricity At Work Regulations, 1989 requirement to earth the external metalwork of all portable tools and equipment, only Class II, or Class III, insulated equipment may be used in industrial situations on SELV supplies.

## Protection by Non-Conducting Location

In a non-conducting location, reliance is placed on the provision of insulating walls and floors and the separation by more than two metres of two exposed conductive parts, or an exposed conductive part and any extraneous conductive part.

This method is intended to prevent simultaneous contact with exposed conductive parts and/or extraneous conductive parts which may be at different potentials following failure of basic insulation.

The non-conducting location must have insulating floors and walls, and the spacing between conducting parts must be at least 2 metres, although this spacing may be reduced to 1.25 metres outside the zone of arm's reach (see Regulation 412 - 05 - 02). Insulating obstacles may be installed to achieve the effective spacing. As an alternative to spacing, extraneous conductive parts may be insulated. Protective conductors must not be introduced into the non-conducting location and it must be ensured that any potential on extraneous conductive parts within the location cannot be transmitted outside the location. Additionally, socket outlets must not incorporate an earthing contact.

This method should be used only in special situations where specified by a suitably qualified electrical engineer. For example, it may be adopted for a test room under supervision adequate to ensure that no change occurs in the design conditions, and where trained persons only are permitted to work.

Special consideration must be given to the design, installation and testing of non-conducting locations (see Inspection and Testing on page 150 ).

## Earth-free Local Equipotential Bonding

In such locations, all exposed conductive parts and extraneous conductive parts which might be accessible at any one time shall be bonded together by an equipotential conductor which must itself not be in contact with earth.

The difficulties inherent in applying the method are obvious. Metal window frames and ceiling suspension channels, if exposed, are among the items to be bonded.

It must be ensured that, in the event of failure of basic insulation, all exposed metalwork in the location will be at the same potential, and conditions for shock will not exist. Thus, the use of this method requires special consideration, and again is only likely to be applied in an industrial testing situation.

## Protection by Electrical Separation

This method may be applied to single items of equipment, such as a shaver unit complying with BS 3052 or, where specified by a suitably qualified electrical engineer, to several items of equipment, in which case the additional rules in Regulation 413 - 06 - 05 must be observed, and the separated installation must be under effective supervision.

The degree of electrical separation must be not less than that between the input and output of a safety isolating transformer complying with BS 3535. The source of supply for the separated circuit may be for example:

   (i) A safety isolating transformer complying with BS 3535.

   (ii) Motor generator having separate windings.

## Residual Current Devices (rcds)

A residual current device is defined in the Regulations as "a mechanical switching device or association of devices, intended to cause the opening of the contacts when the residual current attains a given value under specified conditions". Thus an rcd may be a self-contained piece of equipment, i.e. a residual current circuit breaker (rccb), or it may be a detection/control unit operating a separate circuit breaker, which may be convenient when larger loads require to be protected. An rcd may also incorporate over-current tripping facilities. There are several ways of achieving residual current (previously known as current operated earth leakage) protection as required by the Definition (although the results are the same). It is necessary to know and differentiate between the various types both to secure the desired performance and to avoid damage during testing. It is also necessary to know the parameters against which rcds are designed in order to appreciate what may be expected of them.

All rcds depend upon the action of a current transformer to compare the phase and neutral currents (for single phase), and phases and neutral, if any, for three phase devices. A residual current i.e. one caused by current being conducted by other than the circuit in question, is detected and used to operate the device, opening the associated live conductors. Since rcds must operate on small currents, the initiating power is limited and in some devices is amplified to obtain adequate tripping force. This may be done by self contained electronic devices, powered from the supply passing through the device, or a sensitive relay may operate the trip by an auxiliary supply (see 531 - 02 - 06). It is not always apparent when an rcd incorporates electronic components, and may therefore be sensitive to damage during installation testing. Manufacturers' literature does not always make this clear, and enquiries may have to be made to ascertain whether 500V d.c. testing is permissible. Other rcds depend upon magnetic operation in various forms, and must therefore be situated where they are not subject to stray magnetic fields. (531 - 02 - 07).

It is important to recognize the tolerances in rcd operation laid down in BS 4293. The range is from 50% to 100% of the rated tripping current, e.g. a 30mA device must operate between 15 and 30mA, so that a device which may appear over-sensitive could well meet the requirements of the BS. An rcd to BS 4293 has a maximum breaking capacity of 500A, and a through fault current withstand of around 3kA.

Another point worth noting is that the test push provided on rcds does not necessarily apply only the rated tripping current to confirm operation. In accordance with the BS the test current may be up to 2.5 times the rated tripping current, one reason being that since the test resistor must permit operation on supply voltages less than 240V, when used on 240V the test current must be higher than nominal. Operation of the test push does not prove that the device trips at its rated current, neither does it check the protective conductors or any earth electrode of the installation; it only indicates that the device itself operates with currents of the correct order of magnitude.

Although rcds are designed to give protection against direct and indirect contact, total reliance must not be placed upon them and other recognized methods of protection must still be provided (see Regulations 412 - 01 - 01, 412 - 06 - 01 and 02, and 413 - 01 - 01).

It should be noted that rcds cannot be relied upon to provide discrimination according to their tripping current ratings. This is due partly to the tolerances already mentioned and partly to the fact that fault currents may be of any magnitude, so that if a 30mA rcd is supplied through a 100mA device, they will trip simultaneously with 100mA, and maybe with 50mA; although the smaller device would clear 30mA on its own. If discrimination has to be provided, then it may be done by a time delay on the 'upstream' rcd with the larger current rating. Reference should be made in this connection to the paragraph on 'Minimizing Danger and Inconvenience' on page 49. Regulation 413 - 02 - 19 recommends that the whole of an installation connected to a TT supply be protected by an rcd, and 471- 08 - 06 requires all socket outlets on such a system to be protected by an rcd. To satisfy 314 - 01 - 01 and 02 without providing a main isolating switch connected to two rcds, virtually requires the use of a split-load consumer unit designed for just this purpose or one incorporating a time delayed 100mA 'main' rcd (providing 'Isolating' contact clearances), and with a subsidiary 30mA rcd for the socket outlets.

There are three specific requirements in the Regulations for the provision of rcds and these are:

(1) 471- 08 - 06 requires it for all socket outlet circuits connected to a TT system.

(2) 608 - 13 - 05 requires a 30mA device for supplies to mobile caravans.

(3) 471 - 16 - 01 requires a 30mA device for any socket outlet with a rating of 32A or less likely to be used to supply equipment outside the equipotential bonded area.

**N.B.** A situation of which it is as well to be aware, can occur on installations where a large proportion of the load consists of equipment with solid-state controls, e.g. dimmers, speed controls, and electronic apparatus. These can all give rise to harmonics which distort the basic sine waveform of the supply, and may give rise to something more resembling a pulsating d.c. waveform. If severe enough this can result in tripping only at values higher than the rcd rating, or possibly failure to trip. rcds are available which are designed to operate satisfactorily with such loads. They are more expensive, but the possible need for them should be kept in mind.

**Figure C7**

# Sockets & Fixed Equipment
# Outside Main Equipotential Zone

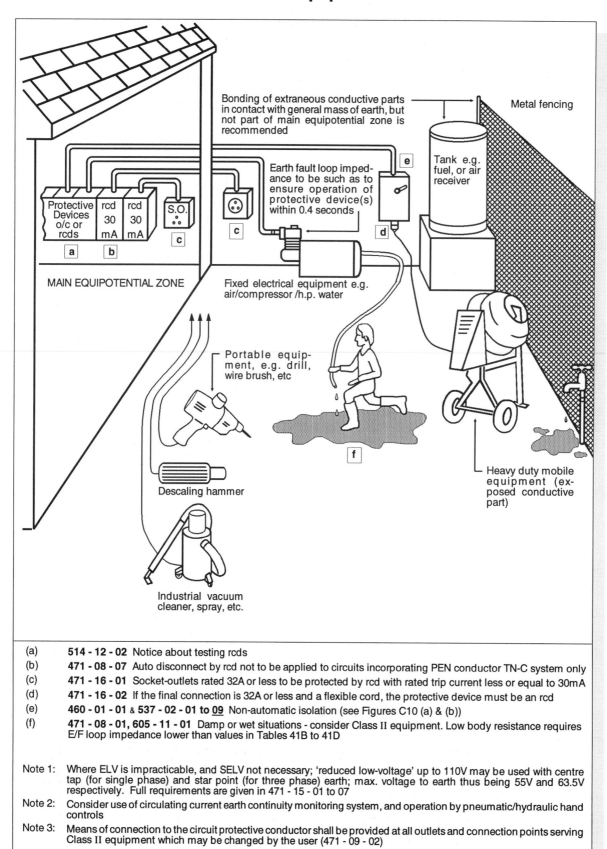

Bonding of extraneous conductive parts in contact with general mass of earth, but not part of main equipotential zone is recommended

Metal fencing

Earth fault loop imped-ance to be such as to ensure operation of protective device(s) within 0.4 seconds

Tank e.g. fuel, or air receiver

Protective Devices o/c or rcds | rcd 30 mA | rcd 30 mA | S.O.

**a** **b** **c**

**c**

**e**

**d**

MAIN EQUIPOTENTIAL ZONE

Fixed electrical equipment e.g. air/compressor /h.p. water

Portable equip-ment, e.g. drill, wire brush, etc

**f**

Descaling hammer

Heavy duty mobile equipment (ex-posed conductive part)

Industrial vacuum cleaner, spray, etc.

(a)   **514 - 12 - 02** Notice about testing rcds
(b)   **471 - 08 - 07** Auto disconnect by rcd not to be applied to circuits incorporating PEN conductor TN-C system only
(c)   **471 - 16 - 01** Socket-outlets rated 32A or less to be protected by rcd with rated trip current less or equal to 30mA
(d)   **471 - 16 - 02** If the final connection is 32A or less and a flexible cord, the protective device must be an rcd
(e)   **460 - 01 - 01 & 537 - 02 - 01 to 09** Non-automatic isolation (see Figures C10 (a) & (b))
(f)   **471 - 08 - 01, 605 - 11 - 01** Damp or wet situations - consider Class II equipment. Low body resistance requires E/F loop impedance lower than values in Tables 41B to 41D

Note 1:   Where ELV is impracticable, and SELV not necessary; 'reduced low-voltage' up to 110V may be used with centre tap (for single phase) and star point (for three phase) earth; max. voltage to earth thus being 55V and 63.5V respectively. Full requirements are given in 471 - 15 - 01 to 07

Note 2:   Consider use of circulating current earth continuity monitoring system, and operation by pneumatic/hydraulic hand controls

Note 3:   Means of connection to the circuit protective conductor shall be provided at all outlets and connection points serving Class II equipment which may be changed by the user (471 - 09 - 02)

**Figure C8**

# Safety Extra-Low Voltage Sources

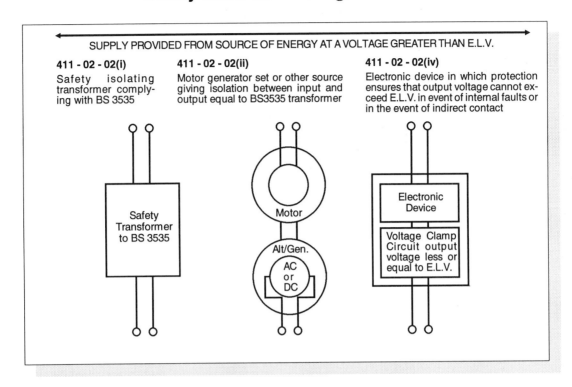

SUPPLY PROVIDED FROM SOURCE OF ENERGY AT A VOLTAGE GREATER THAN E.L.V.

**411 - 02 - 02(i)**
Safety isolating transformer complying with BS 3535

**411 - 02 - 02(ii)**
Motor generator set or other source giving isolation between input and output equal to BS3535 transformer

**411 - 02 - 02(iv)**
Electronic device in which protection ensures that output voltage cannot exceed E.L.V. in event of internal faults or in the event of indirect contact

Safety Transformer to BS 3535

Motor

Alt/Gen.
AC or DC

Electronic Device

Voltage Clamp Circuit output voltage less or equal to E.L.V.

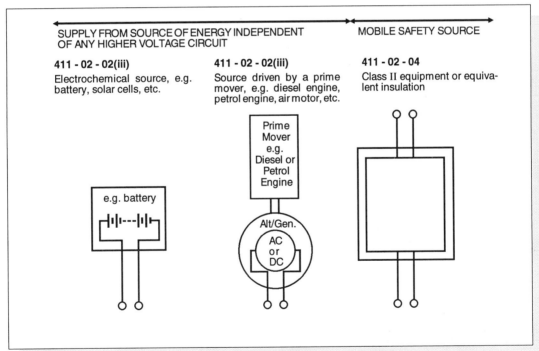

SUPPLY FROM SOURCE OF ENERGY INDEPENDENT OF ANY HIGHER VOLTAGE CIRCUIT

MOBILE SAFETY SOURCE

**411 - 02 - 02(iii)**
Electrochemical source, e.g. battery, solar cells, etc.

**411 - 02 - 02(iii)**
Source driven by a prime mover, e.g. diesel engine, petrol engine, air motor, etc.

**411 - 02 - 04**
Class II equipment or equivalent insulation

e.g. battery

Prime Mover e.g. Diesel or Petrol Engine

Alt/Gen.
AC or DC

**Note:**

Safety Extra-Low Voltage supplies may be obtained from the six types of sources shown. The nominal voltage for such supplies must not exceed 50V rms a.c. or 120V ripple free d.c., and this must be reduced to 25V a.c. (60V d.c.) or less, in situations of unusual risk, e.g. damp or confined spaces (604 - 02 - 02). SELV output supplies and circuits must be isolated from all other conductors and exposed conductive parts. If voltage is between 25V and 50V a.c. (60V and 120V d.c.) then protection against direct contact must be provided by IP2X or IPXXB enclosures or insulation to withstand 500V d.c. test for one minute (411 - 02 - 09)

A bell transformer is only a safety source if it is a Class II safety isolating transformer, double insulated to BS 3535. If it is a Class I or not to BS 3535, then the bell circuit will be anl Extra-low voltage system other than SELV, and must be earthed at one point, any exposed conductive parts being connected to the primary circuit protective conductor. Older bell transformers with cores which accept only single strand conductors with sheath removed contravene Regulation 526 - 03 - 03 and are not acceptable.

For details of SELV and Extra low voltage systems other than SELV see page 54.

# Isolation and Switching

**Isolator.** A mechanical switching device which, in the open position, complies with the requirements specified for isolation. An isolator is otherwise known as a disconnector.

**Isolation.** A function intended to cut off, for reasons of safety, the supply from all or a discrete section of the installation by separating the installation or section from every source of electrical energy.

**Switch.** A mechanical device capable of making, carrying and breaking current under normal circuit conditions, which may include specified operating overload conditions, and also of carrying for a specified time currents under specified abnormal circuit conditions such as those of short-circuit. It may also be capable of making, but not breaking, short-circuit currents.

**Mechanical Maintenance.** Mechanical maintenance is the replacement, refurbishment or cleaning of lamps and non-electrical parts of equipment, plant and machinery.

**Emergency Switching.** An operation intended to remove as quickly as possible, danger, which may have occured unexpectedly.

**Functional Switching.** An operation intended to switch 'on' or 'off' or vary the supply of electrical energy to all or part of an installation.

---

**See Part 2 of BS 7671: 1992 for the complete set of Definitions**

# TOPIC CHART 4 (Information) Isolation and Switching

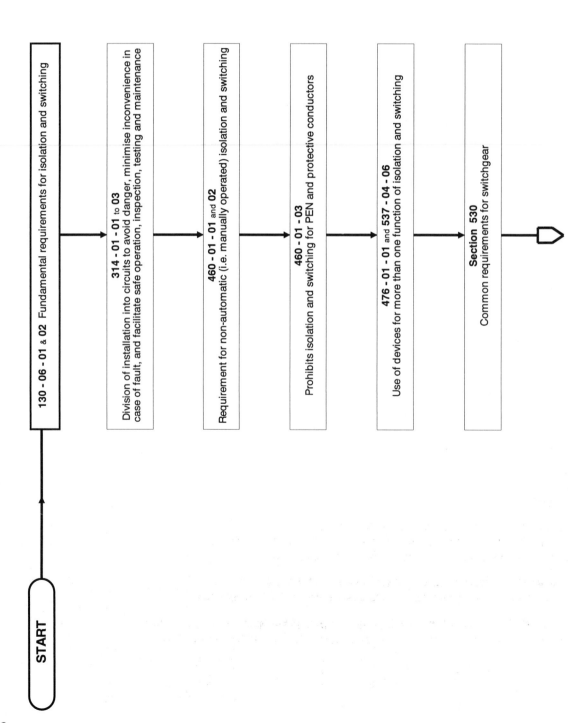

START

**130 - 06 - 01 & 02**  Fundamental requirements for isolation and switching

**314 - 01 - 01** to **03**
Division of installation into circuits to avoid danger, minimise inconvenience in case of fault, and facilitate safe operation, inspection, testing and maintenance

**460 - 01 - 01** and **02**
Requirement for non-automatic (i.e. manually operated) isolation and switching

**460 - 01 - 03**
Prohibits isolation and switching for PEN and protective conductors

**476 - 01 - 01** and **537 - 04 - 06**
Use of devices for more than one function of isolation and switching

**Section 530**
Common requirements for switchgear

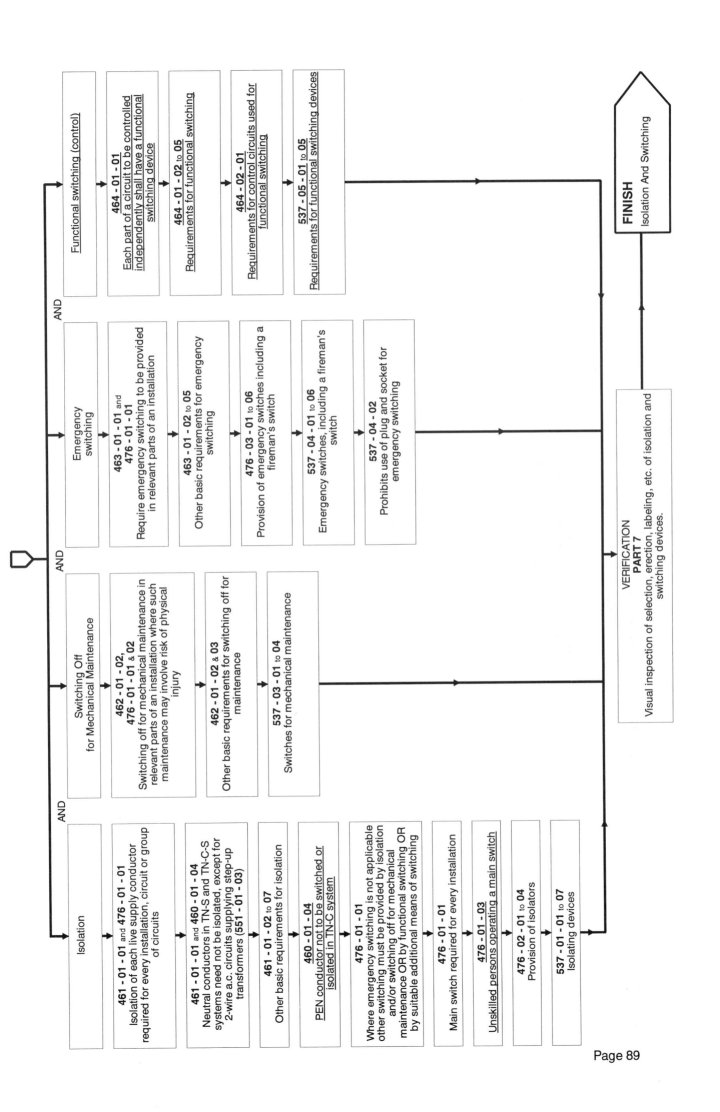

Isolation

**461 - 01 - 01** and **476 - 01 - 01**
Isolation of each live supply conductor required for every installation, circuit or group of circuits

**461 - 01 - 01** and **460 - 01 - 04**
Neutral conductors in TN-S and TN-C-S systems need not be isolated, except for 2-wire a.c. circuits supplying step-up transformers (**551 - 01 - 03**)

**461 - 01 - 02** to **07**
Other basic requirements for isolation

**460 - 01 - 04**
PEN conductor not to be switched or isolated in TN-C system

**476 - 01 - 01**
Where emergency switching is not applicable other switching must be provided by isolation and/or switching off for mechanical maintenance OR by functional switching OR by suitable additional means of switching

**476 - 01 - 01**
Main switch required for every installation

**476 - 01 - 03**
Unskilled persons operating a main switch

**476 - 02 - 01** to **04**
Provision of isolators

**537 - 01 - 01** to **07**
Isolating devices

AND

Switching Off
for Mechanical Maintenance

**462 - 01 - 02,**
**476 - 01 - 01 & 02**
Switching off for mechanical maintenance in relevant parts of an installation where such maintenance may involve risk of physical injury

**462 - 01 - 02 & 03**
Other basic requirements for switching off for maintenance

**537 - 03 - 01** to **04**
Switches for mechanical maintenance

AND

Emergency
switching

**463 - 01 - 01** and
**476 - 01 - 01**
Require emergency switching to be provided in relevant parts of an installation

**463 - 01 - 02** to **05**
Other basic requirements for emergency switching

**476 - 03 - 01** to **06**
Provision of emergency switches including a fireman's switch

**537 - 04 - 01** to **06**
Emergency switches, including a fireman's switch

**537 - 04 - 02**
Prohibits use of plug and socket for emergency switching

AND

Functional switching (control)

**464 - 01 - 01**
Each part of a circuit to be controlled independently shall have a functional switching device

**464 - 01 - 02** to **05**
Requirements for functional switching

**464 - 02 - 01**
Requirements for control circuits used for functional switching

**537 - 05 - 01** to **05**
Requirements for functional switching devices

VERIFICATION
**PART 7**
Visual inspection of selection, erection, labeling, etc. of isolation and switching devices.

**FINISH**
Isolation And Switching

Page 89

## Isolation and Switching

Every installation must be provided with a main switch and a means of isolation, which may be combined in a single device.

Every circuit or group of circuits must be provided with:

(i) a means of isolation (a disconnector);

(ii) a means of interrupting the supply on load; and

(iii) a means of interrupting the supply in any anticipated fault conditions.

The purpose of isolation is to ensure the safety of skilled and/or instructed persons by disconnection of live parts from the supply.

Means of switching off for mechanical maintenance must be provided for every circuit supplying:

– an electric motor

– electrical heating equipment with exposed-to-touch surfaces

– electromagnetic equipment which might cause mechanical accidents

– luminaires (lamp replacement and cleaning are defined as mechanical maintenance)

– any other electrically energized equipment from which possible mechanical or heat hazards could arise from the use of electrical energy.

Emergency switching is required for every part of an installation where it may be necessary to disconnect from the supply rapidly in order to prevent or remove a hazard.

Motors and their starting equipment, individually or in groups, are required to have a means of isolation. Switching for mechanical maintenance and functional switching must also be provided for every motor circuit, and emergency switching must be provided where any electrically driven machine may give rise to danger. Every appliance or luminaire connected other than by means of a plug and socket-outlet, must be controlled by a switch.

Generally, a device may provide more than one of the required means of isolation and switching required by the Regulations, if it is suitable in every respect for each purpose.

## Isolation

– Intended primarily for the use and protection of electrically skilled and/or instructed persons.

An isolator is a device for manual operation which has to be capable of:

(a) opening and closing a circuit under no-load conditions; and

(b) carrying the normal circuit current; and

(c) carrying for a specified time, abnormal currents which may occur during an over-current condition (overload or short-circuit).

The requirements for isolating devices are set out in Regulations 537 - 02 - 01 to 08 and may be summarized as follows:

(i) The device must conform to BS EN 60947-3 in respect of creepage and clearance distances. However, BS EN 60947-3 does not yet contain a requirement for contact separation distance, which has still to be agreed but it is anticipated that the distance will be between 3 and 4mm for low voltage equipment.

(ii) There must be clear indication as to whether the device is in the open or closed position. The indicator must only show that the isolator is open and in the safe position when all the contacts have achieved the specified separation. Alternatively, the position of the contacts must be visible.

(iii) The device must not be capable of being reclosed unintentionally. This requirement may be satisfied either by a feature of its design or the method of its installation.

(iv) An off-load isolating device (isolator) must be capable of being secured against inadvertent or unauthorized operation, for example, by padlocking, placing in a lockable space or enclosure, or by interlocking with a load-breaking device.

Both the definition and the requirements admit the use of switch-fuses, isolating switches and circuit-breakers, provided they meet the contact clearance requirement, but it will be readily apparent that these isolation requirements preclude the use of certain devices such as:

(a) Switches and circuit-breakers (mcbs, rcds, etc.) intended for use on a.c. only, which have less than the specified clearances. "Micro-gap" switches are clearly not acceptable for this purpose.

(b) As stated in Regulation 537 - 02 - 03, no semiconductor device such as a "touch-switch" or "photo-electric switch" may be used for isolation.

The following devices are capable of satisfying the requirements for isolation depending on the particular installation requirements:

(1) Isolators (for off-load disconnection only).
(2) Isolating switches (for on-load disconnection and incorporating required isolating distance).
(3) Links, including fuse-links.
(4) Plugs and socket-outlets.
(5) Circuit-breakers, complying with the contact separation requirement.

The means of isolation provided must disconnect all live conductors – except as noted below. A neutral conductor is (by definition) a live conductor and where a link is inserted it must comply with 537 - 02 - 05.

In no circumstances may means of isolation be provided in a protective conductor, in TN-C systems, the PEN conductor shall not incorporate means of isolation. In TN-S systems, provision need not be made for isolation of the neutral, except where required by Regulation 551 - 01 - 03 dealing with the supply to step-up transformers.

The requirement of Regulation 542 - 04 - 02 for a means of disconnection of the main earthing terminal from the means of earthing may, at first sight, appear to be in conflict with the prohibition of isolation or switching of the protective conductor. This is not so, as the means of disconnection is provided specifically for test purposes when the installation is not in normal use and is under the complete control of electrically-skilled persons. However, when such tests are made, the entire installation associated with the means of earthing must be isolated.

An isolating device which is not required to perform any switching function other than isolation may be an off-load device, in which case it must be capable of being secured against inadvertent or unauthorised operation.

Isolating devices (on-load or off-load) may be installed adjacent to the equipment to be isolated, so that they are under the control of the electrically skilled person working on that equipment. If isolating devices are installed remotely from the equipment to be isolated, provision must be made for securing them against inadvertent reclosure, for example by means of unique keys or non-interchangeable removable handles.

Every isolator must be identified to indicate clearly the circuit with which it is associated.

## Switching off for Mechanical Maintenance

– For use and protection of other than electrically skilled persons.

The devices provided for this purpose are solely to ensure that equipment is safe for persons who are not necessarily electrically skilled to work upon, and no access to live parts is involved or indeed permitted.

Examples of mechanical maintenance are the replacement of lamps and cleaning of luminaires and the cleaning or repair of electrically-driven machines. Where there is possible danger, a means of isolation for mechanical maintenance is necessary to provide additional safety precautions. Examples of such situations are high voltage discharge lighting installations, or where it is necessary to enter or work in close proximity to normally moving parts, such as a rotating drum or a conveyor. The device for this purpose may be inserted in the supply conductors or in the control circuits.

It is preferable to adopt the first alternative, taking care that the device:

    (a) has a reliable or visible indication of the open or off position;

    (b) is so designed or installed as to make a positive action necessary to close it;

    (c) is capable of breaking the full load current of the equipment connected to it; and

    (d) is conveniently situated, readily identifiable and provided with suitable means to prevent unintentional re-closure, caused by e.g. vibration.

In some circumstances, the device may be installed in the control circuits, but such an arrangement requires special consideration to ensure that the degree of safety achieved is not less than that of cutting off the main supply. In general, this will require supplementary precautions such as the use of padlocks or integral locks on the control circuit device. The hazard involved may be burns or physical injury caused by mechanical movement, and warning labels may be required on other control devices such as motor starters, to warn of the presence of two supplies. In such circumstances, a single act of switching off for mechanical maintenance may not be sufficient.

## Emergency Switching

– May be operated by anyone, not necessarily skilled.

As the name clearly indicates, devices provided for this purpose are for cutting off the supply as rapidly as possible to remove a foreseeable dangerous condition. The hazard involved may be electrical (i.e. danger from shock or burns) or mechanical (i.e. danger from movement).

It should be noted that the means provided should act as directly as possible on the supply conductors. The clear implication of this is that direct-acting means should be provided, rather than reliance on other electrical circuits.

A device for emergency switching must be capable of cutting off the full load current of the circuit, including stalled motor current where appropriate.

A push-button which actuates the switch mechanism directly is preferred, although stop-lock or stop-latch push-buttons acting as switches to tripping relays are not prohibited, where they are under the control of the person at risk. Release of such operating devices must not re-energise the equipment unless clear warning is given that this will occur.

Where delays are provided on the closing or opening of contactors, e.g. for undervoltage protection, this must not prevent instantaneous action where necessary for emergency switching.

The consequential effects of emergency switching must be taken into account. For instance, a particular process might involve electric heating elements under which articles pass on a conveyor belt. While it is necessary to provide emergency switching to stop the conveyor belt, it is important that the heating elements should be disconnected at the same time to reduce the risk of fire.

Where there is danger from moving machinery, emergency stopping is necessary. In such situations, the operation of an emergency switch must cause all movement to be rapidly arrested. As well as cutting off the supply, adequate braking arrangements may be needed for high inertia machines and drives.

Obviously, it is necessary to ensure that all devices for emergency switching are accessible and readily identifiable. Attention should be given to this most important point, preferably in consultation with those who are to operate the machinery or plant.

Plugs and socket-outlets are not permitted as a means of emergency switching.

## Functional Switching (Control)

This shall be provided for each circuit or part of a circuit that requires control independently of other parts of the circuit or installation.

This is the term used to denote the normal switching necessary for control of electrically operated equipment of all kinds that may be operated by anyone, not necessarily skilled.

Devices for switching off for mechanical maintenance, or for emergency switching, can also provide switching or control for the normal operation of equipment. If functional switching is not provided in this manner, it is necessary to provide a switch capable of interrupting the on-load supply for any circuit or appliance.

Plugs and socket outlets may be used for functional switching. Plugs and sockets rated at more than 16A must not be used for switching d.c. circuits.

Semi-conductor devices may only be used for functional switching and control subject to compliance with Section 512 of BS 7671: 1992 regarding operational conditions and external influences. They must not be used for any other switching function.

## Combined Devices

Mention was made earlier that it is permissible to combine the functions of isolation, switching off for mechanical maintenance, and emergency switching in one device, provided that the device satisfies all the requirements for each function.

A pole operated fireman's switch in a discharge lighting installation, is an example of a device which may perform the functions of isolation and emergency switching. Depending on its situation in relation to the lighting installation, it may also be suitable for the purpose of mechanical maintenance. However, the requirements for this purpose would not be satisfied if the lighting installation was not reasonably near the switch, and the switch did not have provision for locking-off.

**Figure C9(a)**

# Isolation and Switching

(for identification and notices see Regulations Section 514, and for accessibility see 513)

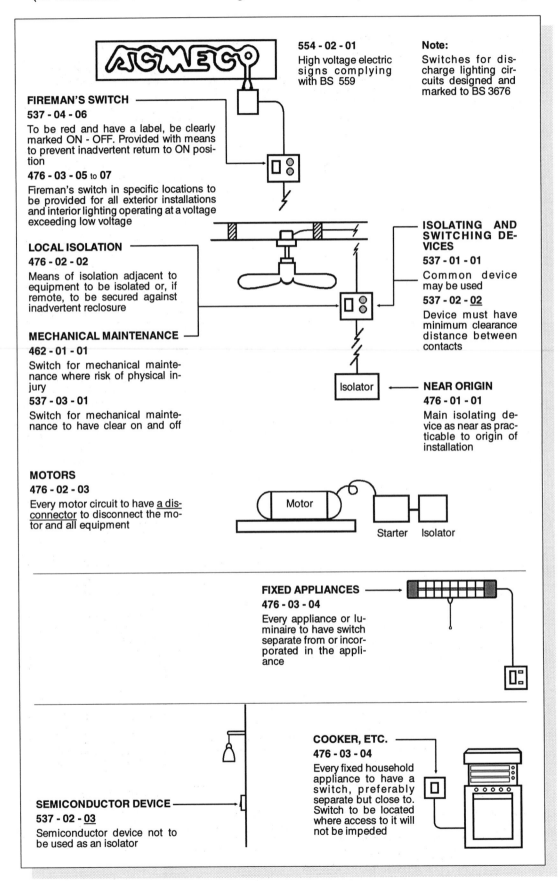

**554 - 02 - 01**
High voltage electric signs complying with BS 559

**Note:**
Switches for discharge lighting circuits designed and marked to BS 3676

**FIREMAN'S SWITCH**
**537 - 04 - 06**
To be red and have a label, be clearly marked ON - OFF. Provided with means to prevent inadvertent return to ON position

**476 - 03 - 05** to **07**
Fireman's switch in specific locations to be provided for all exterior installations and interior lighting operating at a voltage exceeding low voltage

**LOCAL ISOLATION**
**476 - 02 - 02**
Means of isolation adjacent to equipment to be isolated or, if remote, to be secured against inadvertent reclosure

**MECHANICAL MAINTENANCE**
**462 - 01 - 01**
Switch for mechanical maintenance where risk of physical injury

**537 - 03 - 01**
Switch for mechanical maintenance to have clear on and off

**ISOLATING AND SWITCHING DEVICES**
**537 - 01 - 01**
Common device may be used

**537 - 02 - 02**
Device must have minimum clearance distance between contacts

**NEAR ORIGIN**
**476 - 01 - 01**
Main isolating device as near as practicable to origin of installation

Isolator

**MOTORS**
**476 - 02 - 03**
Every motor circuit to have a disconnector to disconnect the motor and all equipment

Motor    Starter    Isolator

**FIXED APPLIANCES**
**476 - 03 - 04**
Every appliance or luminaire to have switch separate from or incorporated in the appliance

**COOKER, ETC.**
**476 - 03 - 04**
Every fixed household appliance to have a switch, preferably separate but close to. Switch to be located where access to it will not be impeded

**SEMICONDUCTOR DEVICE**
**537 - 02 - 03**
Semiconductor device not to be used as an isolator

# Isolation and Switching (continued)

(for identification and notices see Regulations Section 514, and for accessibility see Section 513)

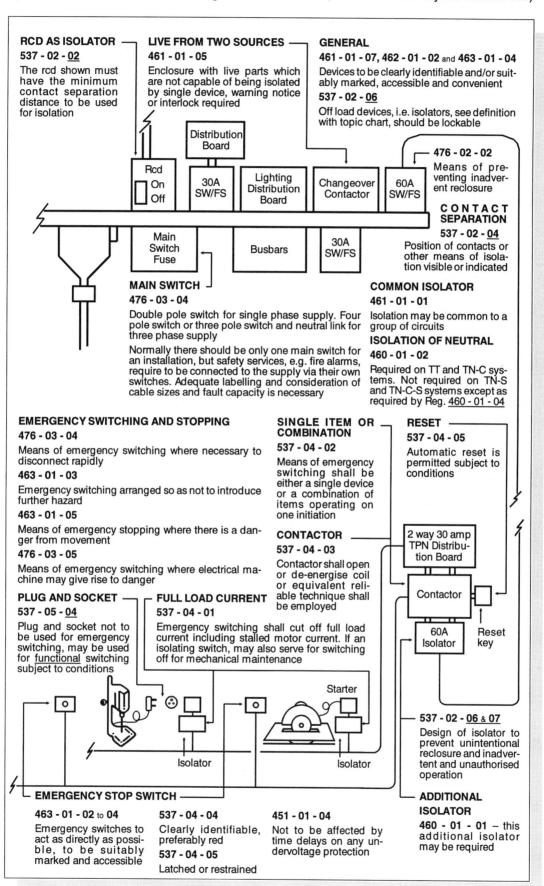

**RCD AS ISOLATOR**
**537 - 02 - 02**
The rcd shown must have the minimum contact separation distance to be used for isolation

**LIVE FROM TWO SOURCES**
**461 - 01 - 05**
Enclosure with live parts which are not capable of being isolated by single device, warning notice or interlock required

**GENERAL**
**461 - 01 - 07, 462 - 01 - 02 and 463 - 01 - 04**
Devices to be clearly identifiable and/or suitably marked, accessible and convenient
**537 - 02 - 06**
Off load devices, i.e. isolators, see definition with topic chart, should be lockable

**476 - 02 - 02**
Means of preventing inadvertent reclosure

**CONTACT SEPARATION**
**537 - 02 - 04**
Position of contacts or other means of isolation visible or indicated

Distribution Board

Rcd On Off

30A SW/FS

Lighting Distribution Board

Changeover Contactor

60A SW/FS

Main Switch Fuse

Busbars

30A SW/FS

**MAIN SWITCH**
**476 - 03 - 04**
Double pole switch for single phase supply. Four pole switch or three pole switch and neutral link for three phase supply

Normally there should be only one main switch for an installation, but safety services, e.g. fire alarms, require to be connected to the supply via their own switches. Adequate labelling and consideration of cable sizes and fault capacity is necessary

**COMMON ISOLATOR**
**461 - 01 - 01**
Isolation may be common to a group of circuits
**ISOLATION OF NEUTRAL**
**460 - 01 - 02**
Required on TT and TN-C systems. Not required on TN-S and TN-C-S systems except as required by Reg. 460 - 01 - 04

**EMERGENCY SWITCHING AND STOPPING**
**476 - 03 - 04**
Means of emergency switching where necessary to disconnect rapidly
**463 - 01 - 03**
Emergency switching arranged so as not to introduce further hazard
**463 - 01 - 05**
Means of emergency stopping where there is a danger from movement
**476 - 03 - 05**
Means of emergency switching where electrical machine may give rise to danger

**PLUG AND SOCKET**
**537 - 05 - 04**
Plug and socket not to be used for emergency switching, may be used for functional switching subject to conditions

**FULL LOAD CURRENT**
**537 - 04 - 01**
Emergency switching shall cut off full load current including stalled motor current. If an isolating switch, may also serve for switching off for mechanical maintenance

**SINGLE ITEM OR COMBINATION**
**537 - 04 - 02**
Means of emergency switching shall be either a single device or a combination of items operating on one initiation

**CONTACTOR**
**537 - 04 - 03**
Contactor shall open or de-energise coil or equivalent reliable technique shall be employed

**RESET**
**537 - 04 - 05**
Automatic reset is permitted subject to conditions

2 way 30 amp TPN Distribution Board

Contactor

60A Isolator

Reset key

Starter

Isolator

Isolator

**537 - 02 - 06 & 07**
Design of isolator to prevent unintentional reclosure and inadvertent and unauthorised operation

**ADDITIONAL ISOLATOR**
**460 - 01 - 01** – this additional isolator may be required

**EMERGENCY STOP SWITCH**
**463 - 01 - 02 to 04**
Emergency switches to act as directly as possible, to be suitably marked and accessible

**537 - 04 - 04**
Clearly identifiable, preferably red
**537 - 04 - 05**
Latched or restrained

**451 - 01 - 04**
Not to be affected by time delays on any undervoltage protection

# Protection against Overcurrent

**Overcurrent** is a current exceeding the rated value for a current-carrying part. It may be either an overload current or a short-circuit current. For conductors the rated value is the current carrying capacity

**Overload Current** is an overcurrent occurring in a circuit which is electrically sound, due to loading in excess of the design current of the circuit.

**Short-circuit Current** is an overcurrent resulting from a fault of negligible impedance between live conductors having a difference in potential under normal operating conditions.

**Earth Fault Current** is a fault current which flows to earth. This may be of a similar magnitude to a short circuit current if the fault between a live conductor and earth is of negligible or low impedance (for Protective Measures, see Section 413).

**Fault** is a circuit conditon in which current flows through an abnormal or unintended path. This may result from an insulation failure or the bridging of insulation. Conventionally the impedance between live conductors or between live conductors and exposed or extraneous conductive parts at the fault position is considered negligible.

**Fault Current** is a current resulting from a fault.

**Prospective Fault Current** is the value of overcurrent at a given point in a circuit resulting from a fault of negligible impedance between live conductors having a difference of potential under normal operating conditions, or between a live conductor and an exposed conductive part.

In the Topic Chart, conditions to be satisfied include the following quantities:-

$I_b$    design current of circuit
$I_n$    nominal current or current setting of protective device
$I_z$    current-carrying capacity of a cable (when installed)
$I_2$    current causing effective operation of the overload protective device
$I$     fault current

---

**See Part 2 of BS 7671: 1992 for the complete set of Definitions**

# TOPIC CHART 5 (Decision) Protection against Overcurrent

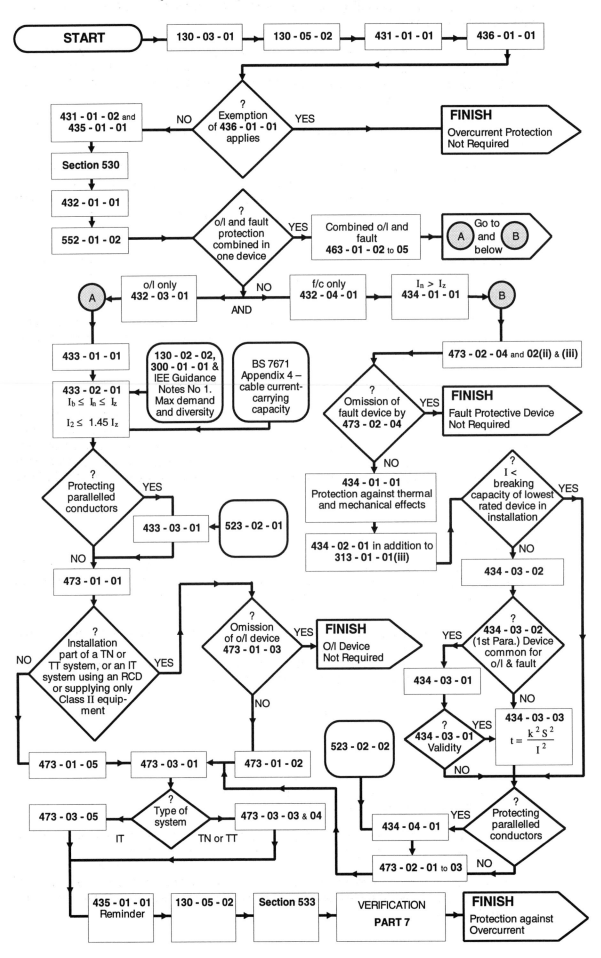

## Protection against Overcurrent

Overcurrent is a current exceeding the rated value for a current-carrying part. It may be an overload current or a fault current.

Overload current occurs in an electrically sound circuit due to the connection of equipment requiring a current in excess of the design current. Short-circuit current flows when a fault of negligible impedance occurs between live conductors which are at different potentials under normal operating conditions.

Three other conditions are possible in addition to overload current and fault current as defined above:

(i) A fault of greater than negligible impedance, (e.g. tracking of insulation) occurring between live conductors, causes a condition similar to overload. Such a condition will either develop rapidly into a fault of negligible impedance and thus cause a short-circuit current to flow, or it will be cleared before this occurs by operation of the overcurrent protective device.

(ii) A fault occurring between a phase conductor and a protective conductor gives rise to an earth fault current of similar magnitude to a short-circuit current, and the overcurrent protective device will operate.

(iii) A fault, not of negligible impedance, between a phase conductor and a protective conductor may pass insufficient current to operate either an overload protective device or a fault protective device.

Conditions (ii) and (iii) are earth fault conditions and are covered in Topic Chart 2.

The overcurrent protective device provided for a particular circuit may not afford protection to the current-using appliance connected to that circuit, or to the flexible cables or cords which may supply it. This must be taken into account in the design of final circuits supplying non-fused socket-outlets. The designer is unlikely to be able to say with certainty what equipment will be connected at any one time. This is a matter to be assessed in accordance with Part 3 of BS 7671: 1992.

## Protection against Overload Current

The rating of the overload protective device is related to the maximum temperature which the insulation of the conductors can withstand for a period of time without significant deterioration. Thus, overload protection is based on two fundamental requirements.

– the design current of the circuit must not be greater than the nominal rating or setting of the related overload protective device;
– the current-carrying capacity of circuit conductors must not be less than the nominal rating or setting of the related overload protective device.

These requirements may be expressed as:-

$$I_b \leq I_n \leq I_z$$

where $I_b$ = design current of circuit;
$I_n$ = nominal current or current setting of a protective device; and
$I_z$ = current-carrying capacity of a cable.

Regulation 433 - 02 - 01 (iii) requires the effective operating current of the device to be not greater than 1.45 times the current-carrying capacity of the related conductor, that is:

$I_2 \leq 1.45 \, I_z$ where $I_2$ is the current ensuring effective operation of the protective device. For all protective devices except semi-enclosed fuses to BS 3036 (see page 109) this requirement is satisfied, if $I_n \leq I_z$.

An overload protective device is capable of carrying indefinitely its nominal current rating or setting, but above this value the speed of operation becomes faster as the current increases. In other words, the time of operation of the device is inversely proportional to the current flowing.

An overload protective device is capable of carrying a small sustained overload current for an appreciable period of time (one hour or more) depending on the rating and type of device, but advantage must not be taken of this when selecting the device. If small overloads are likely to occur frequently, they should be treated as part of the normal design current of the circuit. An example of this is where an electric motor is subjected to frequent starting. This is dealt with later under "Co-ordination of Overload and Fault Protection" (page 104).

## Position of Overload Protective Devices

An overload protective device is normally required at each point along a conductor where its current-carrying capacity is reduced. Such reduction may be caused by a change in cross-sectional area, method of installation, type of cable or conductor, environmental conditions, etc. (see also Topic chart 6 and related notes).

The device protecting the reduced conductor is permitted to be placed along the run on the load side of the point of reduction of $I_z$ , provided that no branches or outlets are introduced between the point of reduction and the device. For example, where a switch-fuse is mounted directly on to a busbar chamber, the conductors connecting the switch-fuse to the busbars may have a lower current rating than the busbars, but are regarded as protected against overload by the device which they are feeding, i.e. the switch-fuse. However, care must be taken that any reduced conductor is also protected against fault current (see below).

Particular attention is drawn to the IEE Guidance Notes No 1 Selection and Erection concerning the connection of a non-fused spur to a ring final circuit. A non-fused spur is permitted to feed only one single or one twin 13A socket-outlet, or one permanently connected load. Clearly, a 30A or 32A overcurrent protective device does not adequately protect a 2.5mm$^2$ cable in accordance with the requirements of Regulations 431 - 01 - 01, 433 - 01 - 01, 433 - 02 - 01, etc.

Acceptance of a non-fused spur having a cable the same size as that used for the ring, depends on the probability that such a spur will never be loaded beyond the current-carrying capacity of the cable. This assumes that the permitted twin 13A socket-outlet will never be subjected, for example, to two 13A loads simultaneously. This installation of non-fused spurs should be avoided, therefore, unless it can be determined that there is no likelihood of the non-protected spur cable being overloaded.

Regulation 473 - 01 - 04(ii) permits the installation of a consumer unit (c.u.) without overload protection of the tails connecting it to the Supplier's meter, provided that the aggregate maximum demand of the c.u. circuits, after any relevant diversity has been applied, is not likely to overload the tails.

## Protection against Fault Current

The thermal effects of fault and short-circuit currents which can result in the melting of insulation and even fire are obvious, but the mechanical effects are often overlooked. Fault currents can produce magnetic fields which distort or break cable cleats and busbar supports. Hence, rapid disconnection from the supply is essential to prevent damage.

Determination of the prospective fault current may be made by measurement or calculation, or it may be ascertained from the Supplier who is required by the Electricity Supply Regulation 1988 as amended, to provide this information, but in no case is it likely to be a simple matter.

Measurement is applicable to energised supplies, and if used to compare with calculated or ascertained values, a useful data base will be acquired.

At the design stage, it will be necessary to calculate values based on the impedances of transformers and mains cables. However, the h.v. source impedance can be ignored and, where conductors of less than 35mm$^2$ csa are involved, the reactances can be disregarded and the calculation made on the resistance values only.

The value calculated is usually that of the prospective short-circuit current at the load side terminals of the overcurrent protective device at the origin of the installation, without any allowance for the impedance of the outgoing circuit or circuits.

The installation designer must use the best information available, and may even need to make his own estimate of the likely range of prospective fault current. Information is included in most manufacturers' literature and in Appendix F of BS 3036 for Semi-Enclosed Fuses.

The Electricity Association Engineering Recommendation P.25, gives details of short circuit characteristics of low voltage distribution networks, with particular reference to 240V single phase supplies up to 100A, with service cutouts having 100A fuses to BS 1361 Type II. Engineering Recommendation P.26 gives similar information for three phase supplies. Reference should be made to these publications (see part E of this Handbook) or to the relevant Supplier.

( P162 )

The following table gives selected values of prospective fault current for service cables and overhead lines up to 25mm$^2$ aluminium or 16mm$^2$ copper connected to supply networks other than those in the area of the London Supplier and in certain large cities, for which the prospective fault current should be assumed to be 16kA at the service cutout, irrespective of the length of the service line.

| Length of Service Line (metres) | Prospective Fault Current kA |
|---|---|
| 8 | 8.8 |
| 15 | 6.0 |
| 22 | 4.5 |
| 26 | 3.9 |
| 40 | 2.7 |

The values given relate to the prospective fault current at the cutout terminals and may be deemed also to apply at the consumers' main switchgear where this is located within approximately 1 metre of the cutout. Where the consumer's switchgear is remote from the cutout, for example in some multi-occupancy buildings, allowance may be made for the additional reduction of the prospective fault current due to the length of sub-main cable.

The length of the service cable may be assumed to be the distance from the cutout to the nearer kerb. (See Engineering Recommendations, P.26.)

Special attention is needed as regards tower blocks of flats where short lateral services may be supplied from substantial rising mains from a basement substation. The prospective fault current at the lower flats will be high but there is little likelihood of an additional substation being installed to increase the prospective fault current at the upper flats, although the original transformer could be increased in capacity to meet a growing demand.

Where the overcurrenct protective device proposed for use at any point in the installation has a breaking capacity less than the prospective fault current at that point, it must be ensured that another protective device having adequate breaking capacity is installed on the supply side. The energy let-through of both devices must be co-ordinated so as to avoid damage to the load side device and to the conductors protected by the devices (see Regulation 434 - 03 - 02).

Although approximate, the prospective fault current for an installation may be determined by the use of a suitable instrument (provided the supply is available when the figure is needed), too much reliance should not be placed on such a reading. If it is low, there may be all the more reason for the supply to be reinforced at a later date, perhaps making switchgear inadequate for the new conditions. The prospective fault current should be the subject of discussion and agreement between the Supplier and the installation designer.

Although the Supplier's cutout fuse is not part of the installation, i.e. it is part of the source of energy, it may generally be used to provide overcurrent protection for the meter tails, and with a consumer unit, the busbar and other connections on the supply side of the overcurrent protective devices in the unit. Under these circumstances, the Supplier may specify a maximum length for the meter tails, or may require additional protection against mechanical damage. The cutout may also provide the back-up protection necessary where the breaking capacity of the fuses or miniature circuit-breakers is less than the prospective fault current at their point of installation. An amendment to BS 5486 Part 13, Consumer Units, permits a manufacturer to allocate a PFC rating to the complete consumer unit when it is preceded by a 100A BS 1361 cutout fuse. For a consumer unit installed in a non-domestic situation, or not preceded by a BS 1361 fuse of 100A or less, then the advice of the manufacturer of the unit should be obtained. The meter tails must be in accordance with Regulation 473 - 02 - 02, i.e. not more than 3m long and enclosed in earthed steel or rigid pvc trunking – if the installation is protected by an rcd as part of a TT system, only rigid pvc trunking may be used.

The requirement that the prospective fault current shall be determined at every relevant point of the installation is automatically satisfied if the prospective fault current at the origin of the installation is less than the breaking capacity of the smallest rated device to be used in the installation.

A point to bear in mind is that the further away from the installation origin an overcurrent protective device is located, the greater will be the impedance of the related conductors, and the lower will be the prospective fault current at the protective device load side terminals. The breaking capacity of the fault protective device can be correspondingly reduced, therefore, but in practice, this may be unnecessary or undesirable, because of a wish to standardise the devices used in an installation and, consequently, their breaking capacities.

Fortunately, the wiring of most installations of a domestic or similar type will reduce the fault level to an acceptable value within a very short distance of the consumer's main switch, so that the probability of the consumer's protective device being subjected to a value of prospective fault current such that back-up protection is required to operate, is very low.

The Regulations permit the overcurrent protective device to be placed other than at the supply end of a cable and a particular fault protective device to be omitted provided that:

(i)   an overcurrent protective device placed at the load end of the cable which affords overload current protection to the cable; and

(ii)   the cable is protected against fault current by another device on its supply side.

This is shown diagrammatically as:

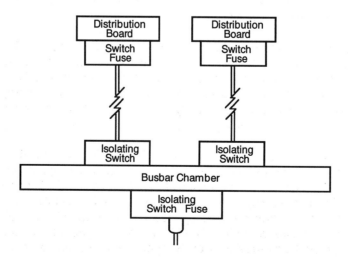

However, the required cable run may be of such impedance that there will be insufficient current at the load end to operate the supply side protective device. In addition, any subsequent variation in the supply network may result in a change in the prospective fault current, which may invalidate the fault protection.

Thus, it will generally be found that the arrangement shown diagrammatically below will be necessary, and the illustration in this section recognises this:

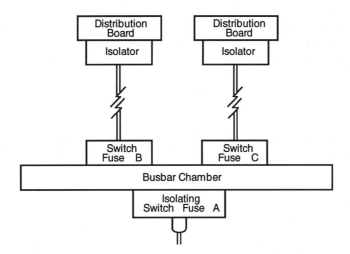

In the above example, devices of lower breaking capacity than the prospective fault current of the whole installation are permitted at points B and C provided that the device at A is suitable. However, it is essential to ensure that should a fault occur on one of the outgoing circuits, the energy let-through of A will not damage the particular device (B or C) or any conductors in the circuit. (Regulation 434 - 03 - 02.)

Overload current protection of the conductors between the busbars and the devices at A and C is dealt with above under Protection against Overload Current. A note on the meaning and application of energy let-through is given below.

Where larger installations and particularly industrial installations are concerned, prospective short-circuit current levels can be 10kA or more, and the rating of equipment used at the intake point must be able to deal with this. Information on the suitability of fault protective devices will usually be available from the manufacturer's literature on a particular device, or directly from the manufacturer when not contained in published information.

Some caution must be exercised if the use of imported equipment is contemplated. The test methods, and the presentation of results obtained, differ widely in various countries and often cannot be compared directly with British practice. Efforts are being made internationally to standardise test methods and parameters.

Where the overcurrent protective device complies with the requirement of Section 433 and has a rated breaking capacity not less than the prospective fault current at the point of installation, it may be assumed that the device is suitable for combined overload and fault protection, but it should be noted that not all circuit-breakers are suitable for this duty – although current-limiting types generally are. The contacts open rapidly, and the arc is dispersed so that both the rate of cooling and the arc's extinction are accelerated.

If a single device is to protect conductors connected in parallel, it is essential that the operating characteristics of the device be chosen carefully, particularly as regards energy let-through, bearing in mind that a fault may occur on one of the conductors. The $I_z$ of such a circuit is the sum of the individual cable ratings (Regulation 433 - 03 - 01). Regulations 433 - 03 - 01 and 523 - 02 - 01 lay down conditions to attempt to ensure equal current sharing between conductors connected in parallel. However, it will still be difficult, if not impossible, to guarantee protection by the use of a single protective device, of each conductor of a number of conductors connected in parallel.

## Co-ordination of Overload and Fault Protection

Attention has already been drawn to the need to co-ordinate the measures for overload and fault protection. This co-ordination is perhaps most familiar in the protection of motor circuits, where overload and fault protection are provided by separate devices, the fault device being located at the origin of the circuit and having an overrated nominal rating but suitable fault characteristic, whilst the overload protective device may be situated in the motor starter. Where, more commonly, a single device provides overload and fault protection, it must have characteristics suited to both functions.

Where an electric motor is subjected to frequent starting, the conductors and overload device must be related to the starting conditions of the motor, not to its running conditions. Infrequent starting of motors can be tolerated by overload protection devices, but careful attention must be given to suitable co-ordination of the characteristics of the associated fault protective device.

The Electricity Association Engineering Recommendation P13/1 states that frequent starting of motors is considered to be starting every two hours or less, but this is related to voltage dip only.

For most situations, advantage can be taken of Regulation 434 - 03 - 02 which allows the use of an overload protective device complying with Section 433 to protect the conductors on the load side of the device, provided that it has a rated breaking capacity not less than the prospective fault current at the point of installation.

## Circuit Limitations

To ensure the protection of conductor insulation against thermal damage during fault conditions (i.e. the energy withstand of the cable, $k^2S^2$, must not be less than the energy let-through of the device, $I^2t$) the requirements are:

(i) for a fuse - the PFC must be sufficiently high (i.e. live conductors must have a correspondingly low impedance) to ensure a short enough disconnection time and

(ii) for a miniature circuit breaker - the same condition applies but, because of the shape of the characteristic curves of these devices, there is also a maximum fault current which must not be exceeded (i.e. live conductors must provide adequate impedance to limit the current to below the breaking capacity of the device). It follows that great care must be excercised if installation and/or modifications require alteration to lengths of conductors protected by miniature circuit-breakers.

## Discrimination and Energy Let-through

Discrimination between all protective devices, whether for overcurrent or shock protection, must be ensured in order to meet the requirements of Regulation 314 - 01 - 02 and 533 - 01 - 06, i.e. to avoid danger and minimise inconvenience in the event of a fault. Discrimination must take into account any fault currents and overload currents which may occur.

The main factor which governs the stress and damage to a conductor is the specific energy let-through of the protective device. The values for a particular device are usually specified in ampere-squared seconds ($A^2s$) and referred to as $I^2t$ let-through.

Manufacturers provide values of $I^2t$ for their devices, and the value for a particular device must not raise the cable conductors to their limiting temperature before disconnection occurs. The method of calculation of the time taken to raise the conductors from the highest permissible temperature in normal duty to limit temperature is given in Regulation 434 - 03 - 03.

The values of energy let-through are also used in assessing discrimination between overcurrent devices in series. In general, if the total energy let-through of the load side device is less than the pre-arcing energy let-through of the supply side device, then discrimination will be obtained. For example, if a fuse-link has a minimum pre-arcing energy let-through of 20kA$^2$s, to obtain discrimination on fault, a down-stream device should have a total energy let-through less than $20kA^2s$.

Manufacturers' advice should be obtained if consideration of $I^2t$ values suggest a less than 2:1 ratio between nominal ratings of series fuses. However, the breaking capacity of any device must be related to the prospective fault current at its point of installation except where advantage is taken of Regulation 434 - 03 - 01, as in the example above (page 102).

Careful assessment of characteristics is necessary to ensure discrimination between fuses and miniature and moulded case circuit-breakers and between circuit-breakers from different manufacturers.

### (a) Fuses

Discrimination is necessary to ensure that a fault will be cleared by the appropriate circuit fuse without affecting other fuses through which it is supplied (Reg 533 - 01 - 06). If fuses have not been selected to achieve discrimination, then although faults will be cleared, it will only be at the expense and inconvenience, at the very least, of replacing more than one fuse or sets of fuses. There will be proper discrimination under the worst conditions, i.e. with the highest attainable fault current, when the total $I^2t$ of the smaller (minor) fuse is less than the pre-arcing $I^2t$ of the larger (major) fuse. Under high fault current conditions fuse operation occurs in less that 0.1 second but with lesser faults taking more than 0.1 second to operate the fuse, reference must be made to the manufacturer's published data for the pre-arcing times of the fuses. The foregoing refers in particular to HBC fuses to BS 88 or 1361, but is still broadly true for semi-enclosed (rewirable) fuses to BS 3036, should their somewhat restricted fault capacity be acceptable for a particular situation. It should be noted that discrimination may fail where, prior to the occurrence of the fault, the major fuse has been heavily loaded, or has deteriorated due to repeated transient overloads.

### (b) Circuit-Breakers

It may be more difficult to ensure discrimination with either miniature or moulded case circuit breakers. This is because at lower (overload) currents, circuit-breakers operate with an inverse time delay due to their thermal mechanisms, but with higher (fault) currents, the magnetic action over-rides the thermal action resulting in almost instantaneous tripping (less than 0.1 second).

| MINATURE CIRCUIT BREAKER TO BS 3871 AND BS EN 60898 – MAIN FEATURES | | | | | |
|---|---|---|---|---|---|
| Type | Nominal rating, $I_n$ (A) | Overload characteristic* | Current band causing instantaneous operation, $I_{inst}$ | Current necessary for instantaneous operation | Common utilisation |
| 1 | $10 \geq I_n$ | $1.50I_n$ | $2.7I_n$ to $4.0I_n$ | $4.0I_n$ | General circuits where load does not exhibit high inrush characteristics, eg. tungsten lighting loads. |
|  | $10 < I_n$ | $1.35I_n$ | | | |
| B | All | $1.45I_n$ | $3.0I_n$ to $5.0I_n$ | | |
| 2 | $10 \geq I_n$ | $1.50I_n$ | $4.0I_n$ to $7.0I_n$ | $7.0I_n$ | General circuits where load only exhibits moderate inrush characteristics, eg. large tungsten and fluorescent lighting loads. |
|  | $10 < I_n$ | $1.35I_n$ | | | |
| C | All | $1.45I_n$ | $5.0I_n$ to $10I_n$ | $10I_n$ | General circuits where load exhibits moderate to high inrush characteristics, eg. motor loads, air conditioning plant etc. |
| 3 | $10 \geq I_n$ | $1.35I_n$ | $7I_n$ to $10I_n$ | $10I_n$ | |
|  | $10 < I_n$ | $1.35I_n$ | | | |
| D | All | $1.45I_n$ | $10I_n$ to $20I_n$ | $20I_n$ | General circuits where load exhibits high to harsh inrush characteristics, eg. X-ray equipment, welding equipment, DOL motors, circuits with transformers, etc. |
| 4 | $10 \geq I_n$ | $1.50I_n$ | $10I_n$ to $50I_n$ | $50I_n$ | |
|  | $10 < I_n$ | $1.35I_n$ | | | |
| * Current required for overload protection within 'conventional time' at reference temperature. Conventional time for Types 1, 2, 3 and 4 and for Types B, C and D up to $I_n \leq$ 63A is 1 hour. For Types B, C and D of $I_n \geq$ 63A the conventional time is 2 hours. | | | | | |

Discrimination can therefore only be attained when the instantaneous tripping current of the smaller circuit- breaker does not approach that of the larger. Account has also to be taken of manufacturing variations in items of the same manufacture, and the different characteristics of other makes and designs. Where circuit-breakers are to be connected in series therefore, the manufacturers, or their data, should be consulted.

(c) Circuit-Breakers with Back-up Fuses

When circuit-breakers are supplied through fuses, sizing should be such that for fault levels up to about 1kA, reliance is placed on the circuit-breaker to clear the fault – and this value will cover the majority of final circuit faults. For higher current levels the fuse will take over, although the change-over point will depend on the current ratio of the two components, and also on the fuse characteristics. However, the use of energy limiting circuit-breakers can reduce the need for fuse back-up, but their shorter operating times may make discrimination difficult, and again, the manufacturer's advice should be sought.

# Live Conductors

**Current-Carrying Capacity of a Conductor** is the maximum current which can be carried continuously by a conductor under specified conditions, without its steady state temperature exceeding a specified value.

**Design Current** (of a Circuit) is the magnitude of the current intended to be carried by the circuit in normal service (rms value for a.c.).

**Live Part** means a conductor or conductive part intended to be energised in normal use, including a neutral conductor, but by convention, not a PEN conductor.

In the Topic Chart, conditions to be satisfied include the following quantities:

$I_b$      = design current of circuit

$I_n$      = nominal current or current setting of a protective device (fuse or overcurrent circuit-breaker)

$I_z$      = current-carrying capacity of a cable (when installed)

$I_t$      = tabulated single-circuit current-carrying capacity of a cable

$C_g$      = correction factor for grouping

$C_a$      = correction factor for ambient temperature

$C_i$      = correction factor for conductors embedded in thermal insulation

$C_t$      = correction factor for operating temperature of conductor ($^{\circ}$C)

$C_d$      = correction factor for semi-enclosed (BS 3036) fuses is 0.725, for other devices, 1.0

---

**See Part 2 of BS 7671: 1992 for the complete set of Definitions**

# TOPIC CHART 6 (Decision) Live Conductors

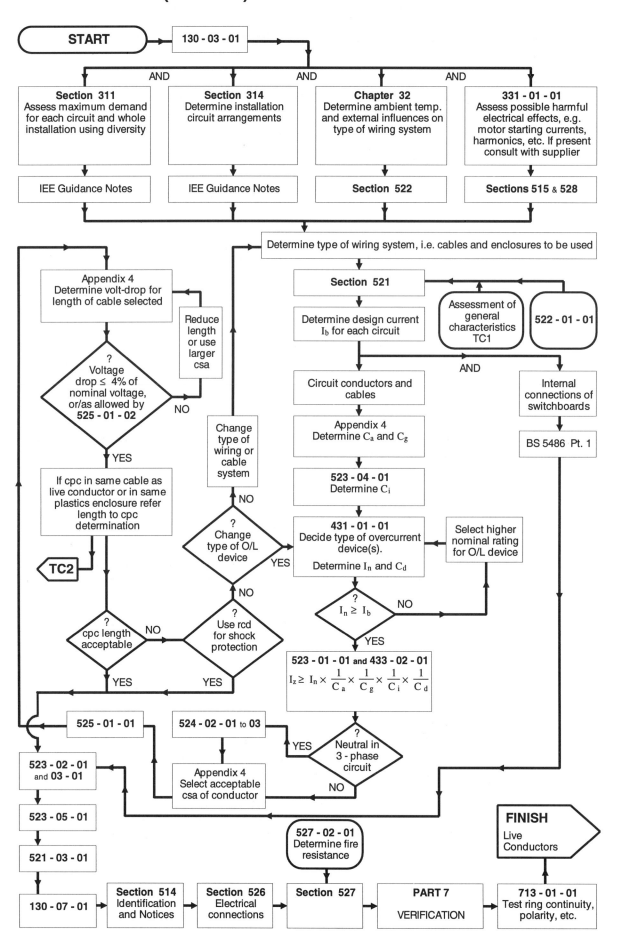

## Live Conductors

The minimum permissible cross-sectional area of the live conductors (phase and neutral) of any circuit depends on:-

– design current of the circuit ($I_b$)
– type of overcurrent protective device and its nominal rating or setting ($I_n$)
– type of cable
– method of installation and environmental conditions
– thermal constraints under fault conditions
– earth fault loop impedance limitations

This chapter deals only with the selection of cables on the basis of current-carrying capacity.

As stated on page 99 of this Handbook, the relationship between $I_b$, $I_n$ and $I_z$ may be expressed as

$$I_b \leq I_n \leq I_z$$

Where an overload protective device has been omitted in accordance with Regulation 473 - 01 - 03 and 04, no value can be assigned for $I_n$ so that the relationship between $I_b$ and $I_z$ becomes

$$I_b \leq I_z$$

For any circuit, it is necessary, therefore, to select a cable having a current-carrying capacity ($I_z$) when installed, which is equal to or greater than $I_n$ or $I_b$ as appropriate.

The values of current-carrying capacity, $I_t$, given in Tables 4D1A to 4L4B of BS 7671: 1992 apply only when a cable is installed under the conditions stated in the headings to the relevant columns of each Table. In those circumstances $I_z = I_t$, and it is, therefore, a simple matter to select a cable having the required current-carrying capacity.

Where the conditions of installation differ from those on which the Tables are based, the effective current-carrying capacity of the cable may vary, thus necessitating the selection of a cable of different cross-sectional area.

The following procedure is recommended for the selection of cables when overload protection is required:

| | | |
|---|---|---|
| 1. | Determine the design current of the circuit | $I_b$ |
| 2. | Select the type and rating of the overload protective device | $I_n \leq I_b$ |
| 3. | Select the type of cable and refer to the appropriate Table 4D1A to 4L4B | |
| 4. | Determine the rating factor for groups, using Table 4B1 | $C_g$ |
| 5. | Determine the rating factor for ambient temperature using either the correction factors in Table 4C1 where semi-enclosed fuses are not used or Table 4C2 where semi-enclosed fuses are used | $C_a$ |
| 6. | Determine the rating factor for thermal insulation (see below) | $C_i$ |
| 7. | If semi-enclosed fuses have been selected and if the cables are not mineral insulated, use factor 0.725. For other types of overcurrent protective device and for mineral insulated cables, the factor is 1.0 | $C_d$ |

8. Determine the tabulated current-carrying capacity -

$$I_t \geq I_n \times \frac{1}{C_g} \times \frac{1}{C_a} \times \frac{1}{C_i} \times \frac{1}{C_d}$$

9. From the appropriate Table 4D1A to 4L2B, select a cable having the value $I_t$ determined at step 8

It should be noted that the effective current-carrying capacity, $I_z$ of the selected cable, when installed, is equal to $\qquad I_t \times C_g \times C_a \times C_i \times C_d$

When overload protection is not required, step 2 is omitted and step 8 becomes:

8. Determine the tabulated current-carrying capacity –

$$I_t \geq I_b \times \frac{1}{C_g} \times \frac{1}{C_a} \times \frac{1}{C_i}$$

## Thermal Insulation

Wherever possible, cables should be routed clear of any space where thermal insulation may be installed. Where this is not possible the current-carrying capacity must be reduced.

Appendix 4 of BS 7671: 1992 gives correction factors for cables installed in accordance with Method 4, Appendix 4.

Where single cables are directly enclosed in thermal insulation for more than 0.5m of their length, their current-carrying capacity shall be taken as 0.5 times the tabulated current-carrying capacity for that cable installed in accordance with Method 1, Appendix 4.

Where the cable is only enclosed in thermal insulation for a short distance, i.e. for less than 0.5m, the correction factors of Table 52A apply.

Where a cable is buried in thermal insulation but is in contact with a thermally conducting surface (e.g. a structural steelwork), the rating factor $C_i$ may be taken from Appendix 4.

Where a cable is installed in accordance with Method 1 of Table 4A, resting on rigid thermal insulation, no further derating is necessary.

## Multiple Application of Rating Factors

When determining current-carrying capacities, it is important to remember that unless the circumstances for reduction in capacity apply simultaneously to a cable, then having derated for the worst situation, no further reduction is necessary. For example, a group of eight circuits of more than one single-core cable running for some distance out from the distribution board would be subject to a grouping factor of 0.5 (Table 4B1). If the cables of these circuits continue either grouped or as a single circuit for more than 0.5m surrounded by thermal insulation, the appropriate factor, 0.5, (Reg. 523 - 04 - 01) would have had to be applied and the resulting multiplier would be 0.25 (i.e. 0.5 x 0.5). However, if only single cables are surrounded by thermal insulation, then the 0.5 grouping factor applied at the start of the circuits already provides the 0.5 derating required by the thermal insulation.

## Thermal Protection of Phase Conductors under Earth Fault Conditions

A situation can arise where an installation apparently in accordance with the Regulations may be lacking in thermal protection of phase conductors during earth fault conditions. This can happen if a circuit is protected against fault, but not overload, at its source, e.g. a distribution board, with overload protection some distance away, e.g. combined with motor control gear. This is a permissible arrangement under 473 - 01 - 01 and 02.

Where fault and overload protection are combined then 434 - 03 - 02 applies. For the situation outlined above, 434 - 03 - 03 is applicable and the adiabatic equation must be satisfied. If the maximum fault current is above that corresponding to the intersection of the adiabatic curve for the phase conductor(s), and the time/current characteristic of the overcurrent protective device, then 434 - 03 - 03 is satisfied, but if the overcurrent is less than this value, the clearance time is greater and the conductors are not thermally protected.

Such a situation can arise when the earth fault current is less than the short-circuit current, caused e.g. by reduced cross-section of protective conductors provided under Table 54G.

With TN-S and TT systems there is every likelihood that earth fault currents will be less than short-circuit currents. In all the circumstances described, the phase conductors as well as the protective conductors must be checked for adequacy against thermal damage.

Attention is therefore drawn to Regulation 543 - 01 - 01 and to Regulation 434 - 03 - 03.

However, where the impedance of the earth return is less than that of the neutral conductor, as may occur with mineral insulated copper covered cables, then the earth fault current can exceed the short-circuit current.

# Chapter 52
## Selection of Types of Wiring Systems

N.B. This illustration does not attempt to indicate all of the specific permissible types
of cables, conductors and material for particular situations referred to in Chapter 52

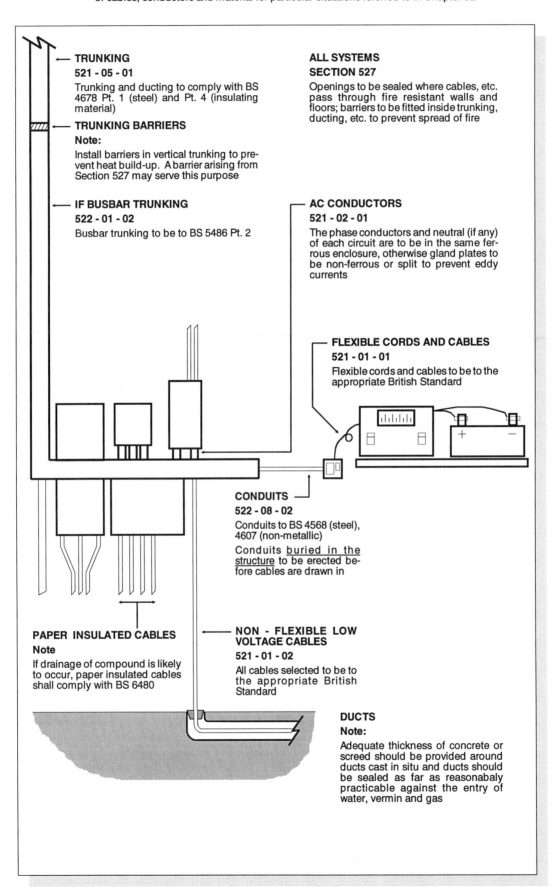

**TRUNKING**
**521 - 05 - 01**
Trunking and ducting to comply with BS 4678 Pt. 1 (steel) and Pt. 4 (insulating material)

**TRUNKING BARRIERS**
**Note:**
Install barriers in vertical trunking to prevent heat build-up. A barrier arising from Section 527 may serve this purpose

**IF BUSBAR TRUNKING**
**522 - 01 - 02**
Busbar trunking to be to BS 5486 Pt. 2

**ALL SYSTEMS**
**SECTION 527**
Openings to be sealed where cables, etc. pass through fire resistant walls and floors; barriers to be fitted inside trunking, ducting, etc. to prevent spread of fire

**AC CONDUCTORS**
**521 - 02 - 01**
The phase conductors and neutral (if any) of each circuit are to be in the same ferrous enclosure, otherwise gland plates to be non-ferrous or split to prevent eddy currents

**FLEXIBLE CORDS AND CABLES**
**521 - 01 - 01**
Flexible cords and cables to be to the appropriate British Standard

**CONDUITS**
**522 - 08 - 02**
Conduits to BS 4568 (steel), 4607 (non-metallic)
Conduits buried in the structure to be erected before cables are drawn in

**PAPER INSULATED CABLES**
**Note**
If drainage of compound is likely to occur, paper insulated cables shall comply with BS 6480

**NON - FLEXIBLE LOW VOLTAGE CABLES**
**521 - 01 - 02**
All cables selected to be to the appropriate British Standard

**DUCTS**
**Note:**
Adequate thickness of concrete or screed should be provided around ducts cast in situ and ducts should be sealed as far as reasonabaly practicable against the entry of water, vermin and gas

# Chapter 52
## Operational Conditions:
## Current-Carrying Capacity

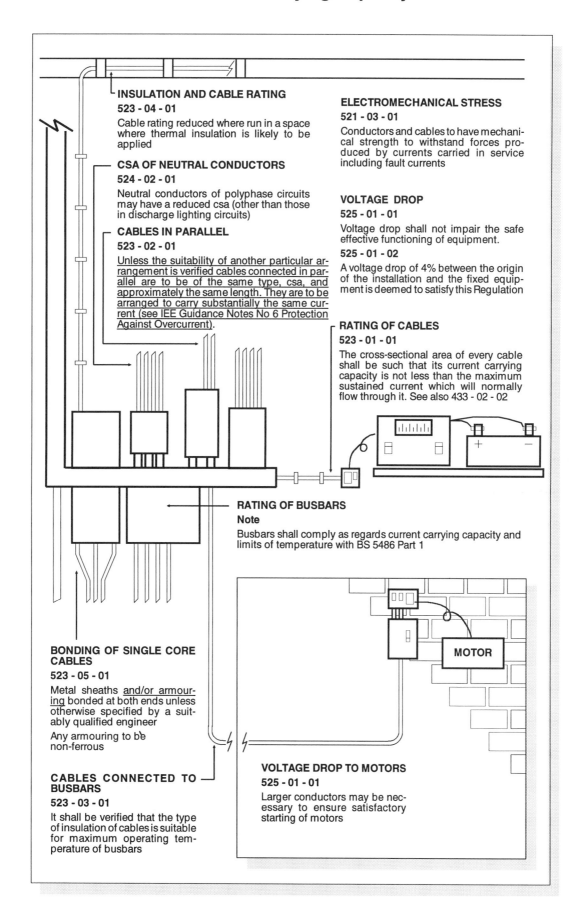

**INSULATION AND CABLE RATING**

**523 - 04 - 01**

Cable rating reduced where run in a space where thermal insulation is likely to be applied

**CSA OF NEUTRAL CONDUCTORS**

**524 - 02 - 01**

Neutral conductors of polyphase circuits may have a reduced csa (other than those in discharge lighting circuits)

**CABLES IN PARALLEL**

**523 - 02 - 01**

Unless the suitability of another particular arrangement is verified cables connected in parallel are to be of the same type, csa, and approximately the same length. They are to be arranged to carry substantially the same current (see IEE Guidance Notes No 6 Protection Against Overcurrent).

**ELECTROMECHANICAL STRESS**

**521 - 03 - 01**

Conductors and cables to have mechanical strength to withstand forces produced by currents carried in service including fault currents

**VOLTAGE DROP**

**525 - 01 - 01**

Voltage drop shall not impair the safe effective functioning of equipment.

**525 - 01 - 02**

A voltage drop of 4% between the origin of the installation and the fixed equipment is deemed to satisfy this Regulation

**RATING OF CABLES**

**523 - 01 - 01**

The cross-sectional area of every cable shall be such that its current carrying capacity is not less than the maximum sustained current which will normally flow through it. See also 433 - 02 - 02

**RATING OF BUSBARS**

**Note**

Busbars shall comply as regards current carrying capacity and limits of temperature with BS 5486 Part 1

**BONDING OF SINGLE CORE CABLES**

**523 - 05 - 01**

Metal sheaths and/or armouring bonded at both ends unless otherwise specified by a suitably qualified engineer

Any armouring to be non-ferrous

**CABLES CONNECTED TO BUSBARS**

**523 - 03 - 01**

It shall be verified that the type of insulation of cables is suitable for maximum operating temperature of busbars

MOTOR

**VOLTAGE DROP TO MOTORS**

**525 - 01 - 01**

Larger conductors may be necessary to ensure satisfactory starting of motors

# Protection against Thermal Effects and External Influences

**Ambient Temperature** is the temperature of the air or other medium where the equipment is to be used.

**External Influence** is any influence external to an installation which affects the design and safe operation of that installation.

**See Part 2 of BS 7671: 1992 for the complete set of Definitions**

# TOPIC CHART 7 (Information) Protection against Thermal Effects and External Influences

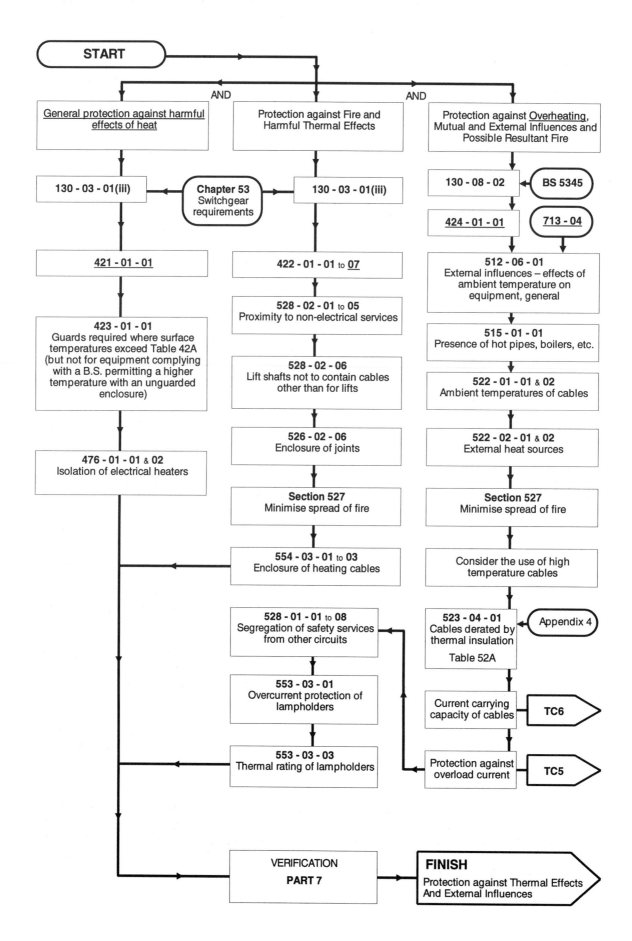

## Protection against Thermal Effects and External Influences

Topic Chart 7 draws together all the Regulations concerned with thermal effects, whether produced by the electrical equipment itself in normal service or due to external causes. In addition, some of the precautions to be taken also assist in restraining the spread of fire through the premises concerned, and notes on these are also included.

**Ambient Temperature** is the temperature of the air or other medium in which equipment is to be used. The ambient temperature of an item of equipment is the temperature at the position where the equipment is to be installed, resulting from the influence of all other equipment in the same location, when operating, but not taking into account the heat generated by the item of equipment itself.

Generally, an item of equipment manufactured in compliance with a British Standard will, where relevant, be specified for operation at a maximum ambient temperature, the design of the equipment being such that the heat generated within it is safely dissipated.

**External Influence** is any influence external to an installation which affects the design and safe operation of that installation. It may affect a particular item of equipment, a circuit, or a whole installation, and may be an influence within or outside a building. For examples of external influences, refer to Section 522 and in particular to codes AA, BE2, CA and CB2, relating to thermal influences.

## Protection against Thermal Effects Produced by Electrical Equipment

The thermal effects dealt with here are those arising from the equipment in normal service. Additional thermal effects caused by overcurrent (Topic Chart 5), overload current and fault current are not considered.

## Protection against Fire

The Regulations identified in the chart under this heading are those intended to minimise the risk of fire caused by the electrical installation.

Regulations 528 - 02 - 01 to 06 define the requirements for the installation of a wiring system in "Proximity to Non-Electrical Services". Compliance will reduce the likelihood of a fire due to a leak from other services in conjunction with the operation of electrical services.

Cables other than those associated with lifts must not run within a lift shaft, not only because of the risk of heat building up in the shaft but also because of the risk of mechanical damage to the cables, and the difficulties associated with inspecting and maintaining them. (528 - 02 - 06).

Regulation 526 - 01 - 01 deals with the soundness of all cable terminations and joints, the obvious reference being to the possibility that a loose or unsound joint could cause an excessive temperature rise, arcing or fire.

Similarly, Regulations 526 - 03 - 02 and 03 refer to the enclosure of terminations or joints in suitable protective material. The requirement for material complying with characteristic "P", BS 476, Part 5. Characteristic "P" indicates that the material has passed the specified test and attention is drawn to the fire terminology.

The fire sealing required under Regulation 527 - 02 - 01 is specifically required to restrict the spread of fire which might otherwise occur through conduits, ducts, etc. An obvious example is a rising main, whether it be comprised of cables or busbars enclosed in metal or cables fixed directly to walls.

A reference is made in Fig. C10 page 112 to thermal barriers which may be required to prevent unacceptable temperatures existing at the top of channel duct, ducting or trunking. These can be the fire barriers, if fitted. Thermal barriers may not have the required degree of fire resistance to act as fire barriers, and if barriers are provided in proprietary rising main equipment, their fire or thermal rating must be ascertained.

# Topic Chart 8

## AN INFORMATION CHART

# Selection and Erection of Equipment

# TOPIC CHART 8 (Information) Selection and Erection of Equipment

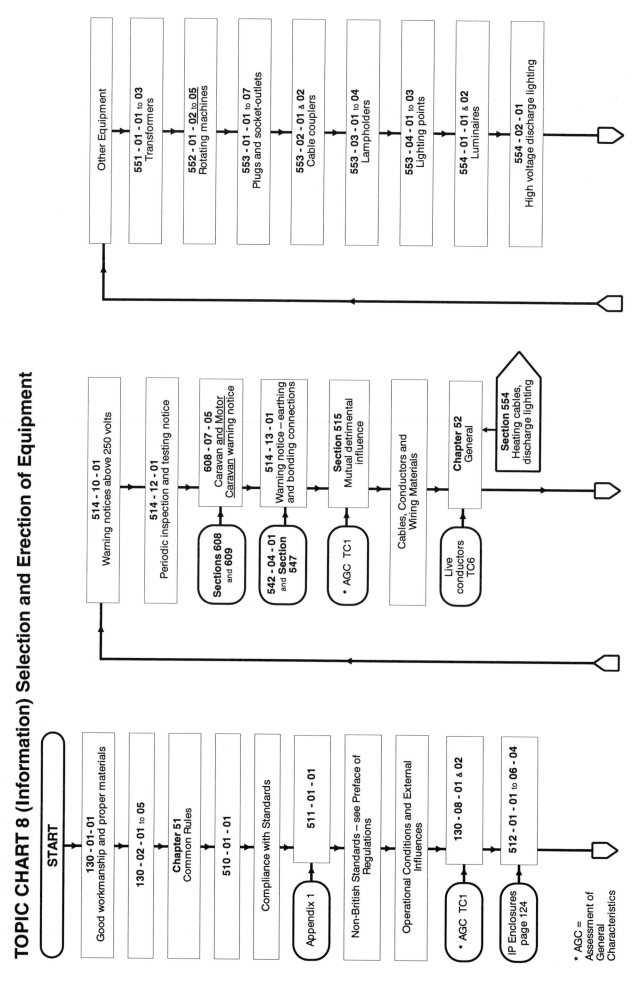

**START**

**130 - 01 - 01**
Good workmanship and proper materials

**130 - 02 - 01** to **05**

**Chapter 51**
Common Rules

**510 - 01 - 01**

Compliance with Standards

**511 - 01 - 01**
→ Appendix 1

Non-British Standards – see Preface of Regulations

Operational Conditions and External Influences

**130 - 08 - 01 & 02**
→ * AGC TC1

**512 - 01 - 01** to **06 - 04**
→ IP Enclosures page 124

**514 - 10 - 01**
Warning notices above 250 volts

**514 - 12 - 01**
Periodic inspection and testing notice

**608 - 07 - 05**
Caravan and Motor Caravan warning notice
→ Sections 608 and 609

**514 - 13 - 01**
Warning notice – earthing and bonding connections
→ 542 - 04 - 01 and Section 547

**Section 515**
Mutual detrimental influence
→ * AGC TC1

Cables, Conductors and Wiring Materials

**Chapter 52**
General
→ Live conductors TC6

**Section 554**
Heating cables, discharge lighting

Other Equipment

**551 - 01 - 01** to **03**
Transformers

**552 - 01 - 02** to **05**
Rotating machines

**553 - 01 - 01** to **07**
Plugs and socket-outlets

**553 - 02 - 01 & 02**
Cable couplers

**553 - 03 - 01** to **04**
Lampholders

**553 - 04 - 01** to **03**
Lighting points

**554 - 01 - 01 & 02**
Luminaires

**554 - 02 - 01**
High voltage discharge lighting

* AGC =
Assessment of
General
Characteristics

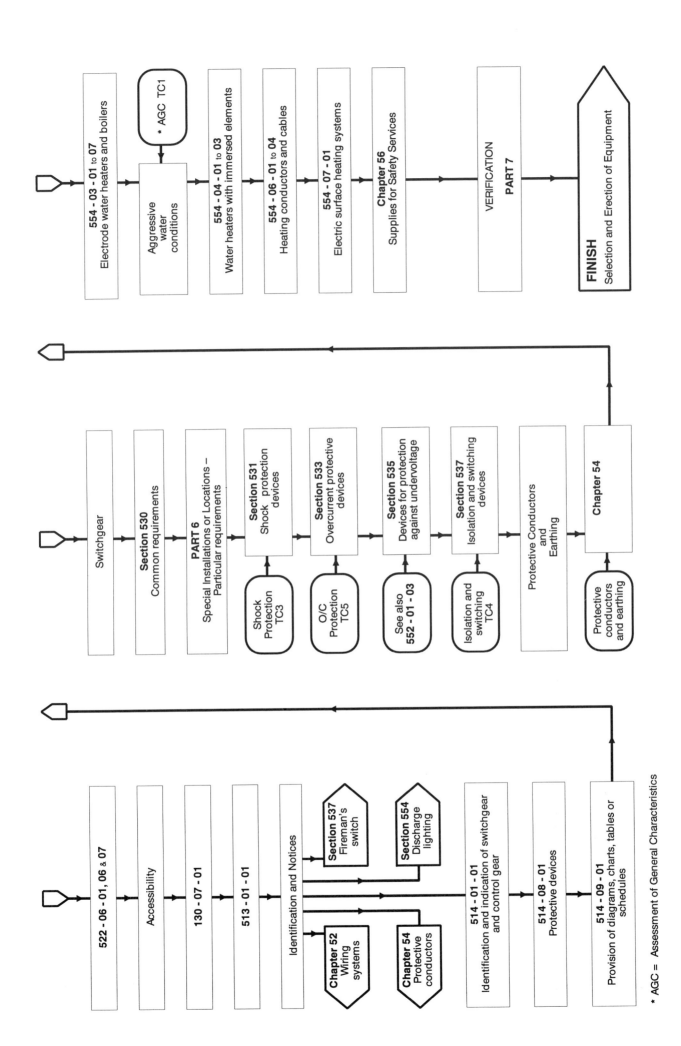

**554 - 03 - 01** to **07**
Electrode water heaters and boilers

* AGC TC1

Aggressive water conditions

**554 - 04 - 01** to **03**
Water heaters with immersed elements

**554 - 06 - 01** to **04**
Heating conductors and cables

**554 - 07 - 01**
Electric surface heating systems

**Chapter 56**
Supplies for Safety Services

VERIFICATION
**PART 7**

**FINISH**
Selection and Erection of Equipment

Switchgear

**Section 530**
Common requirements

**PART 6**
Special Installations or Locations –
Particular requirements

**Section 531**
Shock protection devices

Shock Protection TC3

**Section 533**
Overcurrent protective devices

O/C Protection TC5

**Section 535**
Devices for protection against undervoltage

See also **552 - 01 - 03**

**Section 537**
Isolation and switching devices

Isolation and switching TC4

Protective Conductors and Earthing

**Chapter 54**

Protective conductors and earthing

**522 - 06 - 01, 06** & **07**

Accessibility

**130 - 07 - 01**

**513 - 01 - 01**

Identification and Notices

**Section 537**
Fireman's switch

**Section 554**
Discharge lighting

**Chapter 52**
Wiring systems

**Chapter 54**
Protective conductors

**514 - 01 - 01**
Identification and indication of switchgear and control gear

**514 - 08 - 01**
Protective devices

**514 - 09 - 01**
Provision of diagrams, charts, tables or schedules

* AGC = Assessment of General Characteristics

## Selection and Erection of Equipment

Good workmanship is a requirement of the Regulations, not merely as regards the practical aspects of installation work, but also concerning the fitness of the work being done for its ultimate purpose.

It is anticipated that every installer and designer will need at least access to, if not possession of, the various British Standards referred to by name in the Regulations. Sometimes the information gained from reading the British Standard may seem academic, but the responsibility for the correct selection and use of equipment rests with the designer and installer. Consequently, it is important to note at least the way the various British Standards have been taken into account.

For example, consider BS 4941, which deals with low voltage motor starters. Regulation 435 - 01 - 01 draws attention to the fact that the particular short circuit co-ordination and arrangements of the Standard are not precluded by the Regulation. BS 4941 grades the acceptable damage to starters under faul conditions with an acceptable risk of contact welding. Clearly, the designer or installer ought, at least, to be aware of the possible condition in which equipment may be, following a fault. The performance of the starter should be borne in mind when assessing the importance of the running of the system.

Another example is BS 5266 which incorporates compliance with the Regulations and in particular the volt drop, inspection and testing requirements of emergency lighting installations, all of which are to be taken into consideration by designers and installers.

It is important to bear in mind that any manufacturer may claim compliance with British Standards and, should such a claim be questioned in a particular case, it is necessary to produce evidence that the claim was made by the manufacturer. Thus, it is better (when possible) to use equipment which is marked as to its compliance.

There are certain duties of the Health and Safety at Work Act, Section 6 which are effected by the Consumer Protection Act which concern designers and installers. These are to ensure that all necessary research is done in design or manufacture to eliminate or minimise risk, and to ensure that there is nothing unsafe about the way in which the "article" is erected or installed.

However, the Act does not require the designer/manufacturer/installer to repeat tests or research already done by others, if it is reasonable to rely on the previous tests. As an example, equipment bearing the BSI Safety Mark would probably come under this heading, as far as the inherent safety of that piece of equipment is concerned.

When making a selection of equipment or deciding upon the method of erection protection against damage must be incorporated. This may be achieved by the correct selection of enclosure or by location.

Regulation 522 - 06 - 01 requires that all conductors and cables be protected against mechanical damage to which they may be subject in normal service. This protection can be separate from the conductors, e.g. conduit, or combined with them, e.g. wire armouring. Where cables are concealed within walls or partitions, Regulation 522 - 06 - 06 requires that cables be run within 150mm of the tops of the walls or partitions, or within the same distance of vertical corners. To reach outlets and accessories outside these zones, cables must be run straight, and either vertically or horizontally. Unearthed metallic or plastic capping may be used to protect the cables during plastering or other site operations.

Regulation 522 - 06 - 07 provides that when compliance with 522 - 06 - 06 is impracticable, then either an approved cable having an earthed metallic covering is to be used, or the cable is to be enclosed in earthed conduit, trunking or ducting, which must itself satisfy the requirements for a protective conductor in that situation.

Enclosures are selected according to the protection they give against contact with live conductors either by a part of the body or a conducting material and by the protection they give against the ingress of dust or water. They are known as IP ratings and are detailed on page 124.

In addition the selection must take full account of Section 522 which states the range of external influences to be considered.

The degree of maintainability to be afforded to equipment could lead to dispute, in that the position of equipment might well present difficulties for the maintenance electrician.

Obviously, it is necessary to consider what is reasonable. It may be, for example, that luminaires over a swimming pool require catwalks alongside to facilitate maintenance (usually these would be within a false ceiling space with top access luminaires). Clearly, to have to erect scaffolding spanning the pool would be unreasonable when wishing to clean or maintain the fittings. On the other hand, it may be perfectly reasonable to assume that special access equipment is available for external discharge lighting or advertising signs, but the intention to provide it must be verified.

With the correct choice of equipment sited in an acceptable situation with regard to maintenance, there remains the task of adjusting and co-ordinating the equipment (where such equipment has this facility) to the required degree of protection and the actual installation conditions.

Clearly, the task of final adjustment must carry with it the work of recording the information and characteristics of equipment so adjusted, as well as providing suitable warning notices with respect to the use and purpose of the equipment. This is specifically covered by Regulation 514 - 09 - 01 and in Part 6.

Some information contained within the section of the Regulations concerned with labelling and provision of warning notices is vital to the subject described in the notice, as the requirement is not mentioned elsewhere in the Regulations.

All those concerned with work on building sites will be aware of the problems relating to the final positioning of services, causing routes of pipes, cables, etc., to be modified and redesigned as the building work progresses. With this in mind, it is essential that adequate supervision be given to the installation of cables and conductors, together with frequent verification of the performance of the earth path in terms of impedance values. During the inspection and testing required by Part 7, the adequacy of equipment to disconnect within either 0.4 or 5 seconds must be determined.

Certain specific requirements are made in Chapter 55 under the heading of 'Other Equipment'. Although not a requirement, inductors and transformers should be placed as near as practicable to the associated lamp. This is important for maintenance, and it may also assist in the efficient striking and functioning of the lamp. Mineral insulated cable should not be used to interconnect such lamps and their controlgear, since it can be damaged by the high striking voltages which may be encountered.

Plugs and sockets to BS 1363 and BS 546 shall be used only in locations where the peak ambient temperature will not exceed 35$^{o}$C, and the average temperature will not exceed 25$^{o}$C over 24 hours.

# Index of Protection (IP) Code

| FIRST NUMERAL | | SECOND NUMERAL | |
|---|---|---|---|
| (a) Protection of persons against contact with live or moving parts inside enclosure<br><br>(b) Protection of equipment against ingress of solid bodies | | Protection of equipment against ingress of water | |
| No./SYMBOL | DEGREE OF PROTECTION | No./SYMBOL | DEGREE OF PROTECTION |
| **0** | (a) No Protection<br><br>(b) No Protection | **0** | No Protection |
| **1** | (a) Protection against accidental or inadvertent contact by a large surface of the body, e.g. hand, but not against deliberate access<br><br>(b) Protection against ingress of large solid objects greater than 50mm diameter | **1** | Protection against drops of water. Drops of water falling on enclosure shall have no harmful effect |
| **2** | (a) Protection against contact by standard finger<br><br>(b) Protection against ingress of medium size bodies less than 12mm diameter and greater than 80mm length | **2** | Drip Proof: –<br><br>Protection against drops of liquid. Drops of falling liquid shall have no harmful effect when the enclosure is tilted at any angle up to 15° from the vertical |
| **3** | (a) Protection against contact by tools, wires or suchlike more than 2.5mm thick<br><br>(b) Protection against ingress of small solid bodies | **3** | Rain Proof: –<br><br>Water falling as rain at any angle up to 60° from the vertical shall have no harmful effect |
| **4** | (a) As 3 above but against contact by tools, wires or the like, more than 1.00mm thick<br><br>(b) Protection against ingress of small foreign bodies | **4** | Splash Proof: –<br><br>Water splashed from any direction shall have no harmful effect |
| **5** | (a) Complete protection against contact<br><br>(b) DUSTPROOF: – Protection against harmful deposits of dust, dust may enter but not in amount sufficient to interfere with satisfactory operation | **5** | Jet Proof: –<br><br>Water projected from a nozzle from any direction (under stated conditions) shall have no harmful effect |
| **6** | (a) Complete protection against contact<br><br>(b) DUST TIGHT: – Protection against ingress of dust | **6** | Watertight Equipment: –<br><br>Protection against conditions on ships' decks, etc. Water from heavy seas or power jets shall not enter the enclosures under prescribed conditions |
| IP CODE NOTES<br><br>The degree of protection is stated in form IPXX<br><br>Protection against contact or ingress of water respectively is specified by replacing first or second X digit number tabled, e.g. IP2X defines an enclosure giving protection against finger contact but without any specific protection against ingress of water or liquid<br><br><u>A letter A–D inclusive may be added to the IPXX reference to indicate the degree of protection against access to hazardous parts.</u> | | **7** | Protection Against Immersion in Water: –<br><br>It shall not be possible for water to enter the enclosure under stated conditions of pressure and time |
| | | **8** | Protection Against Indefinite Immersion in Water Under Specified Pressure: –<br><br>It shall not be possible for water to enter the enclosure |

**Note: Use this table for General Guidance only – refer to BS EN60529 for full information on degrees of protection offered by enclosures**

**Figure C12(a)**

# Recommended Methods of Support for Cables, Conductors and Wiring Systems

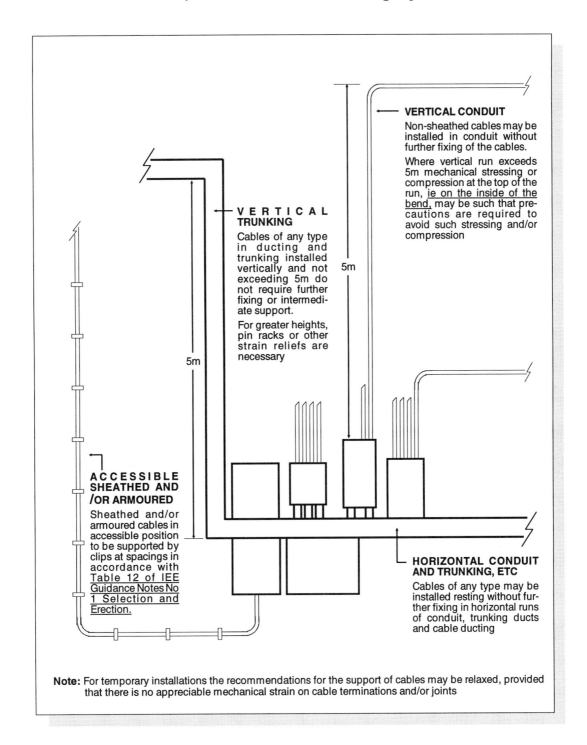

**VERTICAL CONDUIT**

Non-sheathed cables may be installed in conduit without further fixing of the cables.

Where vertical run exceeds 5m mechanical stressing or compression at the top of the run, ie on the inside of the bend, may be such that precautions are required to avoid such stressing and/or compression

**VERTICAL TRUNKING**

Cables of any type in ducting and trunking installed vertically and not exceeding 5m do not require further fixing or intermediate support.

For greater heights, pin racks or other strain reliefs are necessary

**ACCESSIBLE SHEATHED AND /OR ARMOURED**

Sheathed and/or armoured cables in accessible position to be supported by clips at spacings in accordance with Table 12 of IEE Guidance Notes No 1 Selection and Erection.

**HORIZONTAL CONDUIT AND TRUNKING, ETC**

Cables of any type may be installed resting without further fixing in horizontal runs of conduit, trunking ducts and cable ducting

**Note:** For temporary installations the recommendations for the support of cables may be relaxed, provided that there is no appreciable mechanical strain on cable terminations and/or joints

**Figure C12(b)**

# Recommended Methods of Support for
# Cables, Conductors and Wiring Systems (continued)

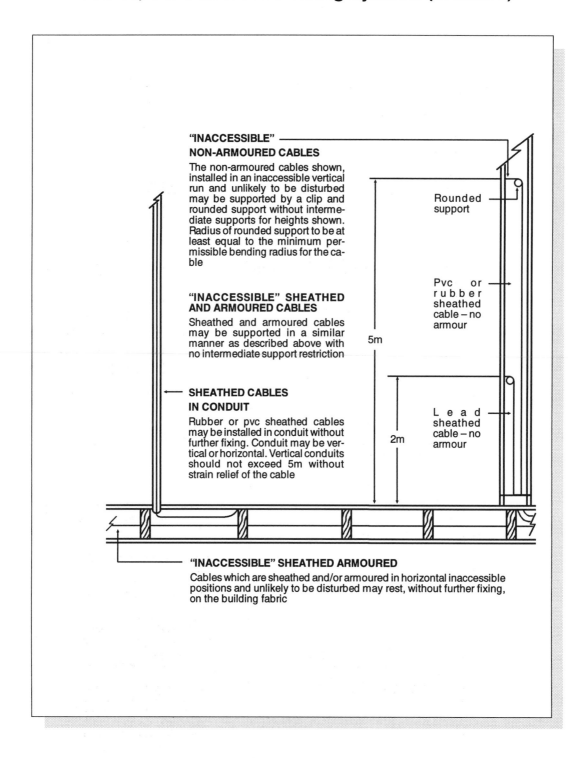

**"INACCESSIBLE" NON-ARMOURED CABLES**

The non-armoured cables shown, installed in an inaccessible vertical run and unlikely to be disturbed may be supported by a clip and rounded support without intermediate supports for heights shown. Radius of rounded support to be at least equal to the minimum permissible bending radius for the cable

**"INACCESSIBLE" SHEATHED AND ARMOURED CABLES**

Sheathed and armoured cables may be supported in a similar manner as described above with no intermediate support restriction

**SHEATHED CABLES IN CONDUIT**

Rubber or pvc sheathed cables may be installed in conduit without further fixing. Conduit may be vertical or horizontal. Vertical conduits should not exceed 5m without strain relief of the cable

Rounded support

Pvc or rubber sheathed cable – no armour

5m

Lead sheathed cable – no armour

2m

**"INACCESSIBLE" SHEATHED ARMOURED**

Cables which are sheathed and/or armoured in horizontal inaccessible positions and unlikely to be disturbed may rest, without further fixing, on the building fabric

# Recommended Methods of Support for Cables used for Overhead Wiring

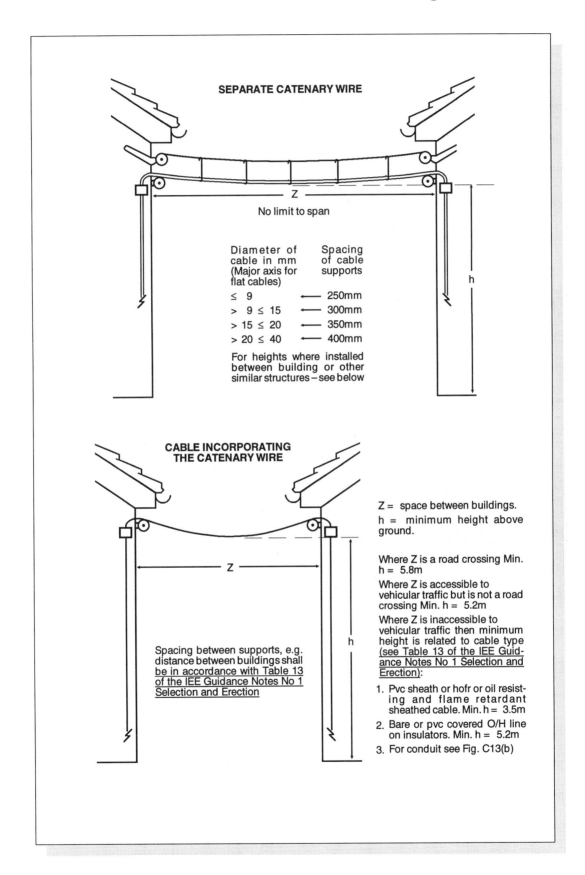

**SEPARATE CATENARY WIRE**

Z

No limit to span

| Diameter of cable in mm (Major axis for flat cables) | Spacing of cable supports |
|---|---|
| ≤ 9 | ⟵ 250mm |
| > 9 ≤ 15 | ⟵ 300mm |
| > 15 ≤ 20 | ⟵ 350mm |
| > 20 ≤ 40 | ⟵ 400mm |

For heights where installed between building or other similar structures – see below

h

**CABLE INCORPORATING THE CATENARY WIRE**

Z

h

Spacing between supports, e.g. distance between buildings shall be in accordance with Table 13 of the IEE Guidance Notes No 1 Selection and Erection

Z = space between buildings.

h = minimum height above ground.

Where Z is a road crossing Min. h = 5.8m

Where Z is accessible to vehicular traffic but is not a road crossing Min. h = 5.2m

Where Z is inaccessible to vehicular traffic then minimum height is related to cable type (see Table 13 of the IEE Guidance Notes No 1 Selection and Erection):

1. Pvc sheath or hofr or oil resisting and flame retardant sheathed cable. Min. h = 3.5m

2. Bare or pvc covered O/H line on insulators. Min. h = 5.2m

3. For conduit see Fig. C13(b)

**Figure C13(b)**

# Recommended Methods of Support for Cables used for Overhead Wiring (continued)

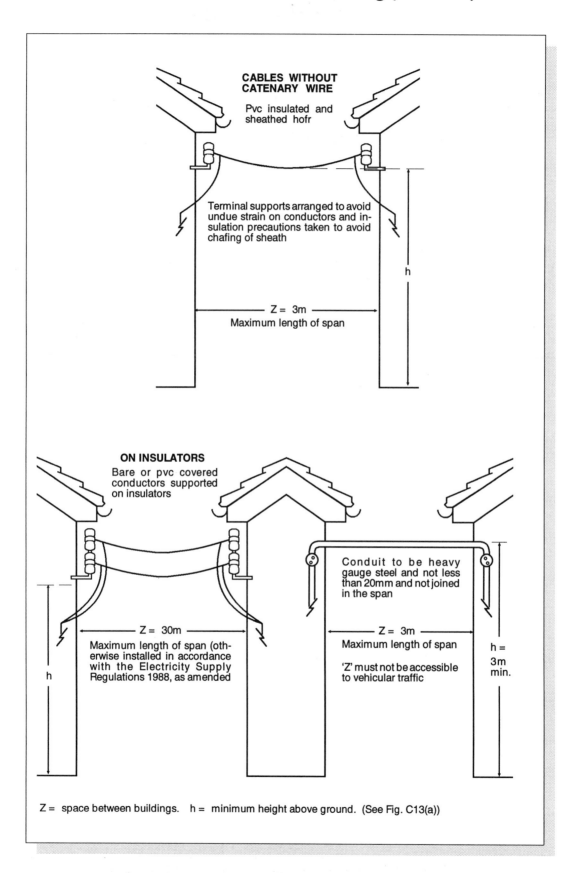

**CABLES WITHOUT CATENARY WIRE**

Pvc insulated and sheathed hofr

Terminal supports arranged to avoid undue strain on conductors and insulation precautions taken to avoid chafing of sheath

Z = 3m
Maximum length of span

h

**ON INSULATORS**

Bare or pvc covered conductors supported on insulators

Conduit to be heavy gauge steel and not less than 20mm and not joined in the span

Z = 30m
Maximum length of span (otherwise installed in accordance with the Electricity Supply Regulations 1988, as amended

Z = 3m
Maximum length of span

'Z' must not be accessible to vehicular traffic

h

h = 3m min.

Z = space between buildings.   h = minimum height above ground.   (See Fig. C13(a))

# Special Installations or Locations – Particular Requirements

**Electrical Installation** is an assembly of associated electrical equipment supplied from a common origin to fulfil a specific purpose and having certain co-ordinated characteristics.

**Restrictive Conducting Location** is a location comprised mainly of metallic or conductive surrounding parts, within which it is likely that a person will come into contact through a substantial portion of his body with the conductive surrounding parts and where the possibility of preventing this contact is limited.

**Highway Power Supplies** is an electrical installation comprising an assembly of associated Highway distribution circuits, Highway distribution boards and Street furniture, supplied from a common origin.

**Street Furniture** is fixed equipment, located on a Highway, the purpose of which is directly associated with the use of the Highway.

**See Part 2 of BS 7671:1992 for the complete set of Definitions**

# TOPIC CHART 9 (Information) Special Installations or Locations – Particular Requirements

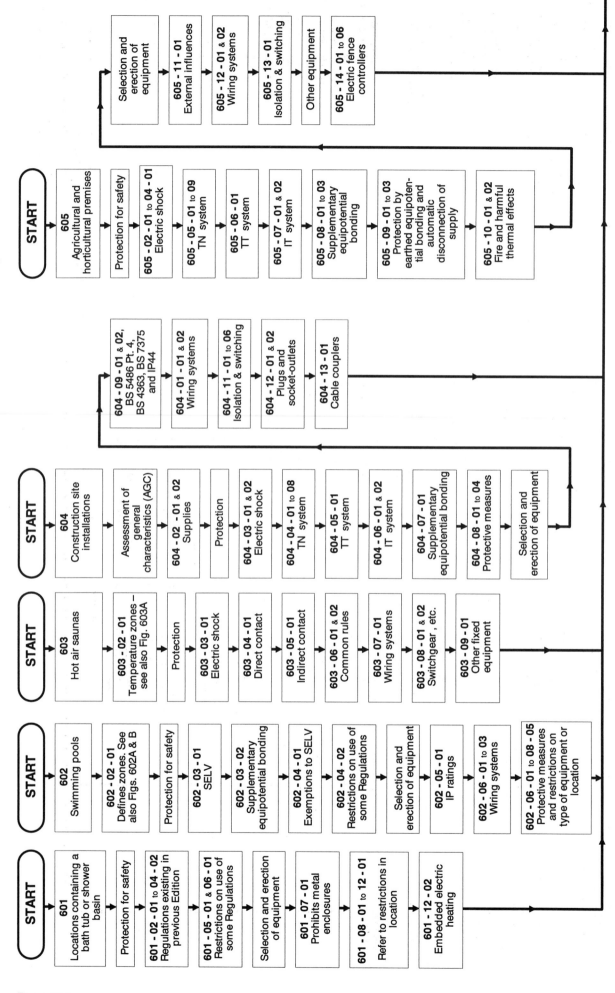

**START**

**601** Locations containing a bath tub or shower basin

Protection for safety

**601 - 02 - 01** to **04 - 02** Regulations existing in previous Edition

**601 - 05 - 01 & 06 - 01** Restrictions on use of some Regulations

Selection and erection of equipment

**601 - 07 - 01** Prohibits metal enclosures

**601 - 08 - 01** to **12 - 01** Refer to restrictions in location

**601 - 12 - 02** Embedded electric heating

---

**START**

**602** Swimming pools

**602 - 02 - 01** Defines zones. See also Figs. 602A & B

Protection for safety

**602 - 03 - 01** SELV

**602 - 03 - 02** Supplementary equipotential bonding

**602 - 04 - 01** Exemptions to SELV

**602 - 04 - 02** Restrictions on use of some Regulations

Selection and erection of equipment

**602 - 05 - 01** IP ratings

**602 - 06 - 01** to **03** Wiring systems

**602 - 06 - 01** to **08 - 05** Protective measures and restrictions on type of equipment or location

---

**START**

**603** Hot air saunas

**603 - 02 - 01** Temperature zones – see also Fig. 603A

Protection

**603 - 03 - 01** Electric shock

**603 - 04 - 01** Direct contact

**603 - 05 - 01** Indirect contact

**603 - 06 - 01 & 02** Common rules

**603 - 07 - 01** Wiring systems

**603 - 08 - 01 & 02** Switchgear, etc.

**603 - 09 - 01** Other fixed equipment

---

**START**

**604** Construction site installations

Assessment of general characteristics (AGC)

**604 - 02 - 01 & 02** Supplies

Protection

**604 - 03 - 01 & 02** Electric shock

**604 - 04 - 01** to **08** TN system

**604 - 05 - 01** TT system

**604 - 06 - 01 & 02** IT system

**604 - 07 - 01** Supplementary equipotential bonding

**604 - 08 - 01** to **04** Protective measures

Selection and erection of equipment

**604 - 09 - 01 & 02**, BS 5486 Pt. 4, BS 4363, BS 7375 and IP44

**604 - 11 - 01 & 02** Wiring systems

**604 - 11 - 01** to **06** Isolation & switching

**604 - 12 - 01 & 02** Plugs and socket-outlets

**604 - 13 - 01** Cable couplers

---

**START**

**605** Agricultural and horticultural premises

Protection for safety

**605 - 02 - 01** to **04 - 01** Electric shock

**605 - 05 - 01** to **09** TN system

**605 - 06 - 01** TT system

**605 - 07 - 01 & 02** IT system

**605 - 08 - 01** to **03** Supplementary equipotential bonding

**605 - 09 - 01** to **03** Protection by earthed equipotential bonding and automatic disconnection of supply

**605 - 10 - 01 & 02** Fire and harmful thermal effects

Selection and erection of equipment

**605 - 11 - 01** External influences

**605 - 12 - 01 & 02** Wiring systems

**605 - 13 - 01** Isolation & switching

Other equipment

**605 - 14 - 01** to **06** Electric fence controllers

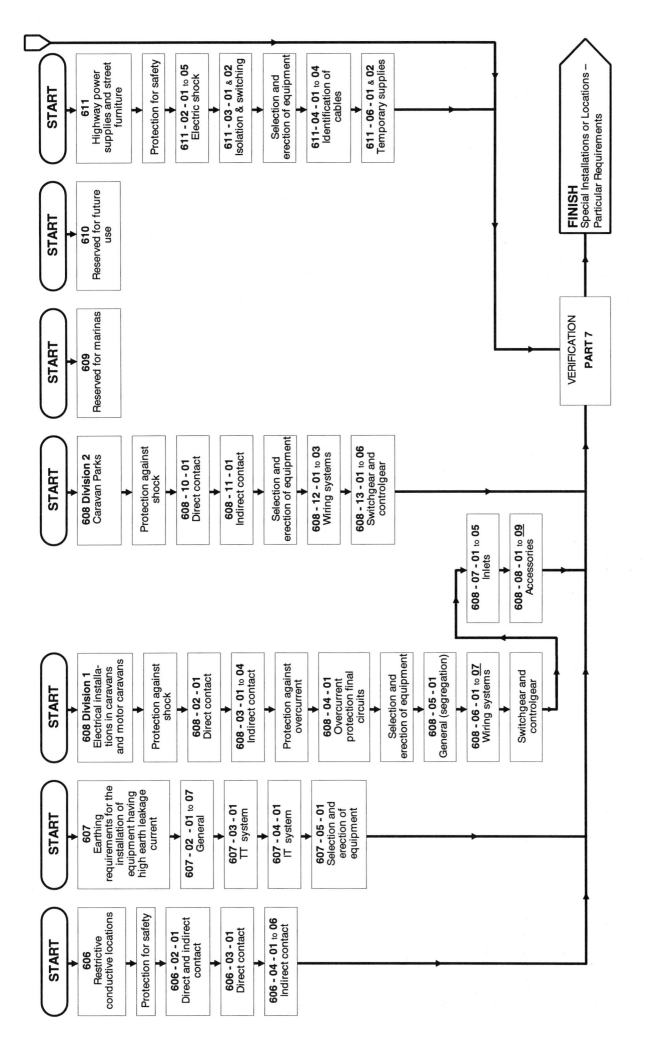

**START**
**611** Highway power supplies and street furniture
Protection for safety
**611 - 02 - 01** to **05** Electric shock
**611 - 03 - 01 & 02** Isolation & switching
Selection and erection of equipment
**611- 04 - 01** to **04** Identification of cables
**611 - 06 - 01 & 02** Temporary supplies

**START**
**610** Reserved for future use

**START**
**609** Reserved for marinas

**START**
**608 Division 2** Caravan Parks
Protection against shock
**608 - 10 - 01** Direct contact
**608 - 11 - 01** Indirect contact
Selection and erection of equipment
**608 - 12 - 01** to **03** Wiring systems
**608 - 13 - 01** to **06** Switchgear and controlgear

**608 - 07 - 01** to **05** Inlets
**608 - 08 - 01** to **09** Accessories

**START**
**608 Division 1** Electrical installations in caravans and motor caravans
Protection against shock
**608 - 02 - 01** Direct contact
**608 - 03 - 01** to **04** Indirect contact
Protection against overcurrent
**608 - 04 - 01** Overcurrent protection final circuits
Selection and erection of equipment
**608 - 05 - 01** General (segregation)
**608 - 06 - 01** to **07** Wiring systems
Switchgear and controlgear

**START**
**607** Earthing requirements for the installation of equipment having high earth leakage current
**607 - 02 - 01** to **07** General
**607 - 03 - 01** TT system
**607 - 04 - 01** IT system
**607 - 05 - 01** Selection and erection of equipment

**START**
**606** Restrictive conductive locations
Protection for safety
**606 - 02 - 01** Direct and indirect contact
**606 - 03 - 01** Direct contact
**606 - 04 - 01** to **06** Indirect contact

VERIFICATION **PART 7**

**FINISH** Special Installations or Locations – Particular Requirements

Page 131

## Particular Requirements for Special Installations or Locations

The additional requirements for special installations or locations are set out in Part 6 under the following headings:

| | |
|---|---|
| 600 | General |
| 601 | Locations containing a bath tub or shower basin |
| 602 | Swimming pools |
| 603 | Hot air saunas |
| 604 | Construction site installations |
| 605 | Agricultural and horticultural premises |
| 606 | Restrictive conducting locations |
| 607 | Earthing Requirements for the installation of equipment having High Earth Leakage Currents |
| 608 Division 1 | Electrical installations in Caravans and Motor Caravans |
| 608 Division 2 | Electrical Installations in Caravan Parks |
| 609 | (Reserved for Marinas) |
| 610 | (Reserved for future use) |
| 611 | Highway power supplies and street furniture |

"Highway Power Supplies and Street Furniture", Section 611; this section relates to places used by the public and designated as a Highway. Similar equipment in areas used by the public which are not part of a building and which are not designated a Highway, must also comply with Section 611 requirements, e.g. pedestrian precinct or market square closed to traffic.

**Figure C14**

# Locations Containing a Bath Tub or Shower Basin

## FOR BONDING REQUIREMENTS SEE FIGURE D6

**DISCONNECTION TIME**

**601 - 04 - 01**

For circuits supplying equipment where such equipment is simultaneously accessible with exposed conductive parts or with other equipment or with extraneous conductive parts. A 0.4 second disconnection time must be achieved

**STATIONARY APPLIANCE**

**601 - 12 - 01**

No stationary appliance of which the heating elements can be touched to be installed within reach of person using bath or shower

**601 - 07 - 01**

No surface metallic enclosures; exposed earthing and bonding conductors

**601 - 02 - 01**

No electrical equipment to be installed in the interior of a bath tub or shower basin

**ELECTRIC SHAVER**

**601 - 09 - 01**

Shavers only connected through unit to BS 3535 and earth terminal of unit to be connected to its circuit protective conductor

**NO SOCKET OUTLETS – except SELV**

**See below**

**601 - 10 - 02**

There shall be no socket outlets and no provision for connecting portable appliances in rooms containing a fixed bath or shower (other than shaver and SELV sockets – see below)

**601 - 10 - 01**

12V a.c. or d.c. SELV socket outlets may be installed in bathrooms out of reach of a person using the bath if the socket has no accessible metal parts and protection against direct contact with live parts in accordance with 411 - 02 - 09

**LAMPHOLDERS**

**601 - 11 - 01**

Parts of a lampholder within a distance of 2.5m from the bath or shower shall be constructed of, or shrouded in, insulating material. Bayonet type (B22) lampholders to have shield to BS 5042 (HO Skirt) or, as an alternative, totally enclosed luminaires may be used

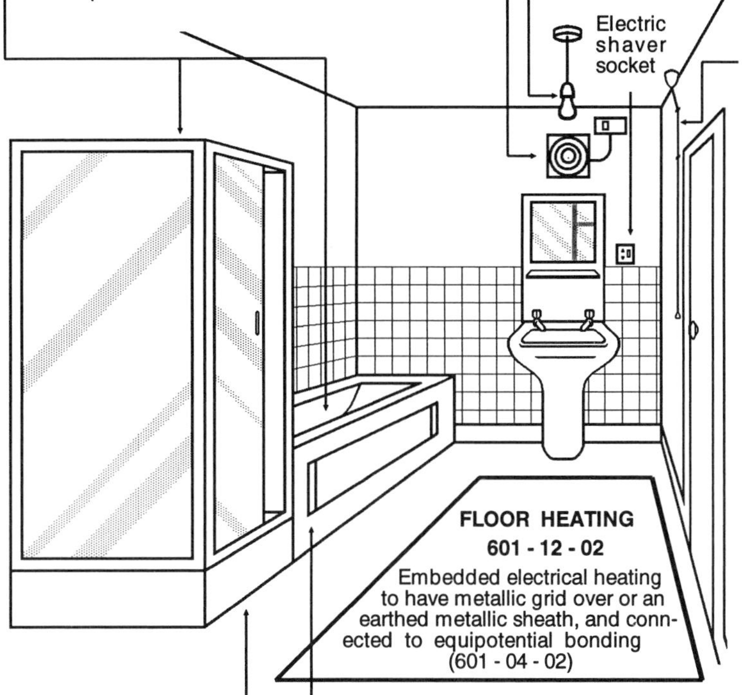

Electric shaver socket

**SWITCHES AND CONTROLS**

**601 - 08 - 01**

Every switch, electrical control or adjustment to be situated so as to be normally inaccessible to a person using the bath or shower – this does not apply to:

(i) insulating cords of switches

(ii) mechanical actuators with linkages incorporating insulating components, of remotely operated switches

\* (iii) controls incorporated in water heaters <u>and shower pumps</u> complying with <u>British Standards</u>.

(iv) Switches supplied by SELV

(v) shaver supply units to comply with 601 - 09 - 01

\* Notwithstanding (iii) above, it is recommended that a local cord-operated switch be provided for maintenance purposes. Cord operated switches which also provide isolation must be double pole, with adequate contact separation and clear mechanical indication of the switch position given only when both poles have achieved full separation. (Neon indicators do not suffice.) Alternatively, switches may be situated outside the bath or shower room

**FLOOR HEATING**

**601 - 12 - 02**

Embedded electrical heating to have metallic grid over or an earthed metallic sheath, and connected to equipotential bonding (601 - 04 - 02)

**SHOWER CUBICLE**

**601 - 10 - 03**

Where shower cubicles are located in rooms other than locations containing a bath tub or shower basin, any socket outlets shall be situated at least 2.5m from the cubicle

**601 - 04 - 03**

Bath panel to need a tool to remove it if electrical equipment installed behind

**Figure C15**

# Caravans:
## Methods of Support for Cables, Conductors and Wiring Systems
### (Cables in Particular Conditions)

**LUMINAIRES**
**608 - 08 - 06**
Luminaires should preferably be fixed directly to the caravan structure, but if pendants are fitted there must be a means of securing the luminaire against damage caused by movement of the caravan

**FIXING OF CABLES**
**608 - 06 - 05**
Unless enclosed in rigid conduit, cables shall be supported at least every 0.4m vertically and 0.25m horizontally

**CARAVAN AND MOTOR CARAVAN INLETS**
**608 - 07 - 01** to **03**
Caravan connectors and inlets to be to EN 60309 - 2, two pole and earth. Inlet marked with nominal voltage and be in protected external position on the caravan.

**FUEL STORAGE**
**608 - 06 - 06**
No electrical equipment (including cables) shall be installed in any compartment intended for the storage of gas cylinders

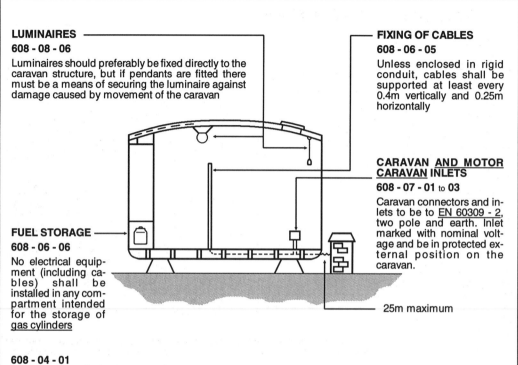

— 25m maximum

**608 - 04 - 01**
All final circuits to have o/c protection which disconnects all live conductors

**N.B.** The above relates to caravans of all types, i.e. residential and touring. Regulation 608 - 07 - 05 requires a notice to be fixed near the main switch in touring caravans. For clarity, this requirement is stated in this illustration and in Figure C15. The notice requires each rcd to be tested every time the caravan is connected to a supply.

# Caravans, Motor Caravans, their Pitches and Caravan Parks

**(For an individual maximum of 16A demand)**

## ALL SYSTEMS

### PROTECTIVE CONDUCTORS

**608 - 03 - 02**

Where protection against indirect contact is afforded by automatic disconnection of supply the protective conductor to be installed throughout each circuit within the van, sockets in caravan to incorporate earthing contact

### EXTRANEOUS PART BONDING

**608 - 03 - 04**

Extraneous conductive parts to be bonded to the caravan protective conductor. If construction of van does not ensure continuity, multiple bonding may be used

Minimum size of bonding conductor to be 4mm$^2$

### WARNING NOTICES

**608 - 07 - 04**

In all touring caravans, a notice (specified in this Regulation) to be fixed near the main switch inside the van. See also Regulation 514 - 12 - 02 and Figure C14

A notice concerning the periodic testing of RCDs is required at the main switchboard

### CARAVAN PITCH FIXED SOCKET (a)

**608 - 13 - 01 & 02**

Sockets to be to EN 60309-2 220/240V, IPX4 minimum 16A and positioned within 20m of the caravan intake position and individually overcurrent protected

**608 - 13 - 05**

PME terminals are not to be connected to socket outlets supplying caravans. Protective conductors at sockets are to be connected to an independant earth electrode, i.e. to form a TT system

### EARTH FAULT PROTECTION (b)

**608 - 13 - 05**

Each van to be supplied from a socket outlet having 30mA rcd protection. Up to 3 vans may be supplied from one rcd

## TN - S SYSTEM

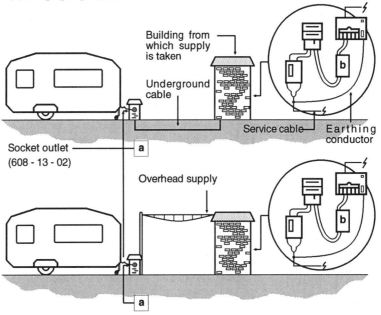

Building from which supply is taken

Underground cable

Socket outlet (608 - 13 - 02)

Service cable

Earthing conductor

Overhead supply

When the pitch is fed by an o.h. supply a duplicated protective conductor is required or connection to the main earthing terminal by protective conductor unlikely to fail

## TT SYSTEM

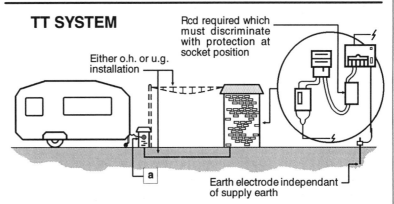

Rcd required which must discriminate with protection at socket position

Either o.h. or u.g. installation

Earth electrode independant of supply earth

## TN - C - S SYSTEM

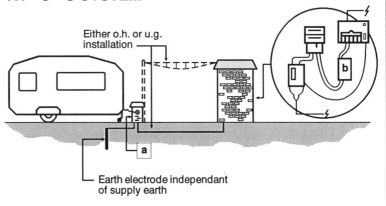

Either o.h. or u.g. installation

Earth electrode independant of supply earth

**(a)** Rcd and/or overcurrent protective device

**(b)** Position of rcd when protecting up to 3 caravans

# Inspection and Testing

BS 7671:1992 differentiates between:

**Initial verification - for new work, additions and alterations.**

AND

**Periodic Inspection and Testing**

However the test equipment and testing methods are the same for both.

TC10 has been segregated to reflect these three separate subjects.

### TC10A - Inspection and Testing for Initial Verification

Regulation 711-01-01 requires testing and inspection "during erection" to overcome some of the problems arising from having to disconnect equipment such as luminaires if testing is carried out after completion.

This Regulation also recognises that installations can only be verified so far as is "reasonably practicable" that they meet the requirements of the regulations.

On new installations certain information (see Regulations 514-09-01 and 711-01-02) must be readily available and if it is not, the requirements of BS 7671:1992 cannot be satisfied.

On complex installations it is necessary to establish at the outset the types of installation present (e.g. SELV, Functional Extra-Low Voltage, non-conducting locations, etc.) and the system employed (see page 000).

It is also be necessary to confirm that the installation complies with Part 3 of the Regulations relating to Assessment of General Characteristics (AGC). Information regarding prospective fault current and external earth fault loop impedance will also be required.

### TC10B - Periodic Inspection and Testing

With an existing installation, the problems may well be complicated by lack of information or by additions and alterations for which no drawings or information exist. A reasonably practicable approach must be taken to all periodic inspection and testing.

An existing installation then, should be examined to establish, among other things, whether SELV systems or extra-low voltage systems other than SELV are present. Delineation of equipotential zones should be ascertained. This may seem a simple matter, but, what may be simple to identify on a drawing, may be very difficult to appreciate on site. Equipotential zones which, on a cursory inspection, may appear to require a connection with earth, could turn out to have been designed as earth-free equipotential zones where the provision of an earth connection would, under fault conditions, lead to dangerous voltage on simultaneously accessible equipment. Furthermore,it may be that what appears to be an earth-free equipotential zone is really an equipotential zone designed to be earthed, the earth connection of which has been lost through unauthorised work or other reasons.

Clearly then, the need to establish how or if it was intended to satisfy BS 7671:1992 requirements when the installation was initially designed involves identifying what exists. Since what exists may not be all of what was originally intended, the responsibility on the person carrying out the inspection and testing is onerous.

However complex these problems may seem, they can usually be solved by a logical approach. Situations could occur, however, where there is no alternative but to accept that it is impossible to resolve the state of an installation without wholesale redesign and consequent re-construction.

### TC10C - Methods of Testing and Test Instruments

This section deals with all the requirements of test instruments for both initial verification and periodic inspection and testing.

# TOPIC CHART 10A (Decision) Inspection and Testing for Initial Verification

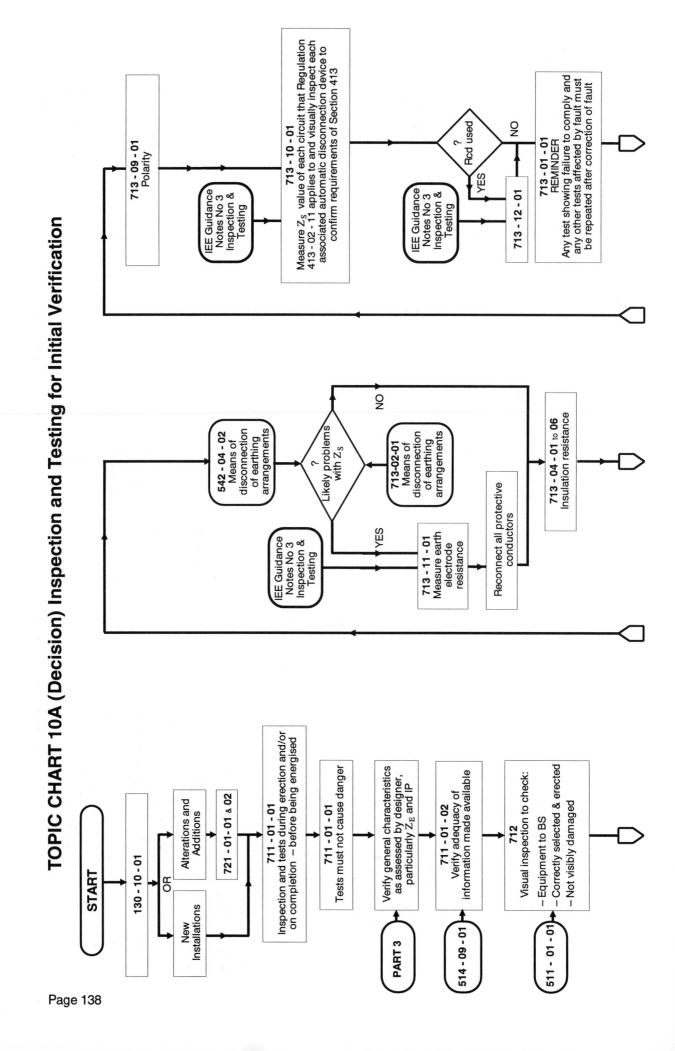

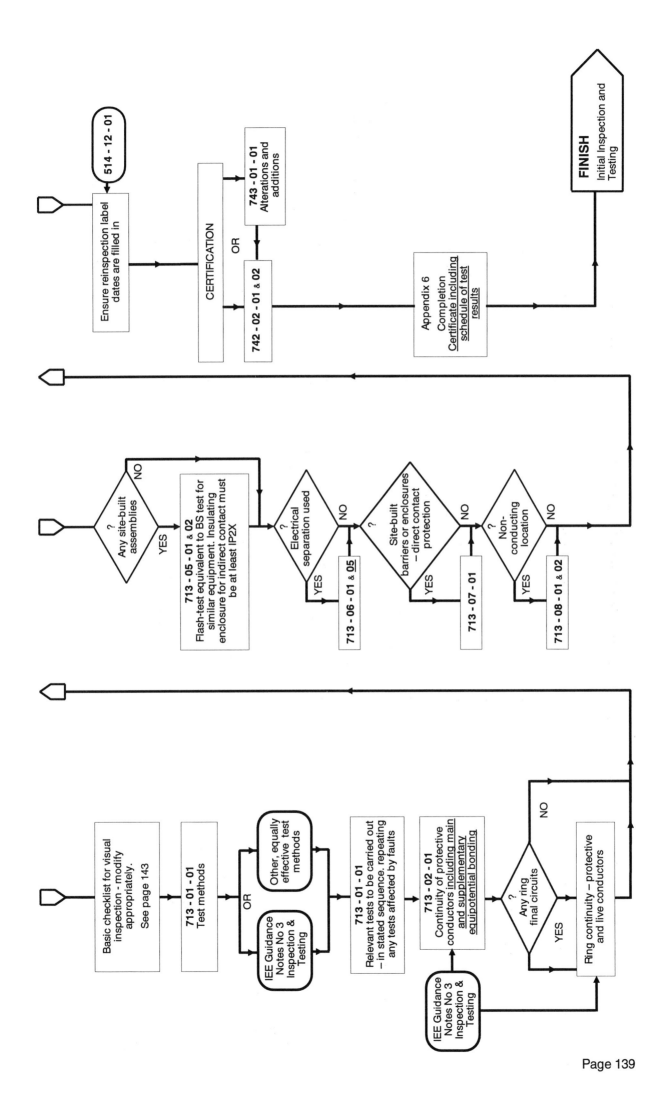

**514 - 12 - 01**

Ensure reinspection label dates are filled in

**CERTIFICATION**

OR

**743 - 01 - 01**
Alterations and additions

**742 - 02 - 01 & 02**

Appendix 6
Completion Certificate including schedule of test results

**FINISH**
Initial Inspection and Testing

---

Any site-built assemblies? — NO

YES

**713 - 05 - 01 & 02**
Flash-test equivalent to BS test for similar equipment. Insulating enclosure for indirect contact must be at least IP2X

Electrical separation used? — NO

YES

**713 - 06 - 01 & 05**

Site-built barriers or enclosures – direct contact protection? — NO

YES

**713 - 07 - 01**

Non-conducting location? — NO

YES

**713 - 08 - 01 & 02**

---

Basic checklist for visual inspection - modify appropriately. See page 143

**713 - 01 - 01**
Test methods

OR

Other, equally effective test methods

IEE Guidance Notes No 3 Inspection & Testing

**713 - 01 - 01**
Relevant tests to be carried out – in stated sequence. repeating any tests affected by faults

**713 - 02 - 01**
Continuity of protective conductors including main and supplementary equipotential bonding

IEE Guidance Notes No 3 Inspection & Testing

Any ring final circuits? — NO

YES

Ring continuity – protective and live conductors

Page 139

## Inspection

Inspection must be carried out before any testing is undertaken.

Regulation 712 - 01 - 03 lists the items which must be checked during the inspection. However, this list is not exhaustive, and may be supplemented or adapted for particular purposes, such as the inspection of conduit or trunking before the cables are drawn in. It is also necessary to check that the cables installed are correct relative to the nominal current and/or setting of protective devices. This list, reproduced with suitable columns in which notes or comments can be made against each item, should be available to all those responsible for inspection and testing.

The requirement that the visual inspection should include a check of the choice and setting of monitoring and protective devices (712 - 01 - 03(xii)) is a demanding responsibility because, for example, the check must include reference to the characteristics of the devices mentioned in Regulation 413 - 02 - 04, and the method used to comply with this Regulation, which will determine the required maximum earth fault loop impedance.

Obviously, the information referred to above is of considerable value to persons responsible for re-inspections and testing, and must always be noted on the Completion Certificate. It is also recommended that the Completion Certificate be retained with the record drawings and any other particulars of the installation.

## Design Information

Some of the Regulations allow a degree of flexibility in the way compliance with a fundamental requirement is achieved, and in order to confirm compliance, it is necessary that those responsible for inspection and testing understand fully the design requirements and their basis. In particular, attention is drawn to the detailed information required on the Completion Certificate (BS 7671:1992 Appendix 6). The designer of the installation must provide this information. It may also be necessary to ascertain how the designed cross-sectional areas of protective conductors, etc., have been determined.

Some aspects of design are based on the prospective fault current at the origin of the installation. This value should be checked by calculation or by measurement.

## Verification of Protective Devices

The sequence of measurements necessary to verify the selection and co-ordination of devices for overcurrent and shock protection is shown below.

1. Verify circuit design current ($I_b$)

2. Determine overcurrent protective device type and rating ($I_n$)

3. Determine shock protection device if different from 2. Determine operating characteristics of any rcd

4. Determine type of cable and rating factors, e.g. $C_a$, $C_g$, $C_i$, $\underline{C_t}$ (Appendix 4)

5. Divide $I_n$ by correction factors = $I_t$

6. Determine csa of conductor having tabulated current-carrying capacity (app.4) which is the next larger than $I_t$

7. For verification of overcurrent protection go to Box 18

8. Determine volt-drop at furthest point of circuit

9. (i) Measure $R_1$ and $R_2$ at furthest point of the circuit applying correction factor from Appendix G $= \dfrac{(R_1 + R_2)}{C}$
   of IEE Guidance Notes No 1 Selection and Erection if temperature differs from $20^{\circ}$C.
   (ii) Measure $Z_e$
   (iii) Select multiplier M from Appendix G of IEE Guidance Notes No 1 Selection and Erection according to cable insulation type
   (iv) Determine $Z_s = Z_e + \dfrac{M}{C_a}(R_1 + R_2)$ (value of $C_a$ from IEE Guidance Notes No 3)

10. Ascertain effective csa of protective conductors

11. If protective conductor effective csa satisfies Table 54G then 543 - 01 - 01 is satisfied. Go to box 14

12. If protective conductor csa does not comply with Table 54G, then the csa of the protective conductor must be established according to 543 - 01 - 03

13. For socket outlet circuits 413 - 02 - 08 is satisfied if:
    (a) $Z_s$ (Box 9 above) is less than Table 41B1(fuses) or 41B2(mcbs), or,
    (b) $Z_s$ (Box 9 above) is less than Table 41D, and M times $R_2$ is less than Table 41C value

14. For fixed equipment circuits, 413 - 02 - 09 is satisfied if $Z_s$ (Box 9 above), is less than Table 41D value

15. Where an rcd provides shock protection, the value of $Z_s$ (Box 9), must not exceed 50 divided by the rated operating current of the rcd
    It must be determined that the relevant disconnection times are met

16. Where an rcd provides shock protection the installation must also satisfy the requirements of 543 - 01 - 01 for minimum csa of protective conductor and the requirements for protection against thermal effects

17. Establish the operating current of the overcurrent protective device compared with lowest current-carrying capacity cable which it protects against overcurrent

18. If the design (load) current $I_b$ is less than or equal to the nominal current $I_n$ of the device AND:

19. If the nominal current $I_n$ does not exceed the lowest current-carrying capacity $I_t$, after all the correction factors are applied then:

20. Compliance with 433 - 02 - 01 is ensured

21. Where overload device is also used for fault protection its maximum fault current breaking capacity must be compared with the prospective fault current then:

22. If breaking capacity is equal to or not less than the prospective fault current measured at the point at which the device is installed then compliance with Regulations is generally ensured with regard to fault protection for cables on load side

23. Where conditions mentioned in Box 22 do not apply it is necessary to check that the time of interruption of supply does not allow the cables to exceed the limit temperature (see 434 - 03 - 03)

24. Energy let-through of separate fault protective devices must be compared with the maximum permissible energy for the next load-side fault protective device
    Devices must be co-ordinated so that the maximum permissible let-through is not exceeded

All appendix references are to BS 7671:1992

# TOPIC CHART 10B(Decision) Periodic Inspection and Testing

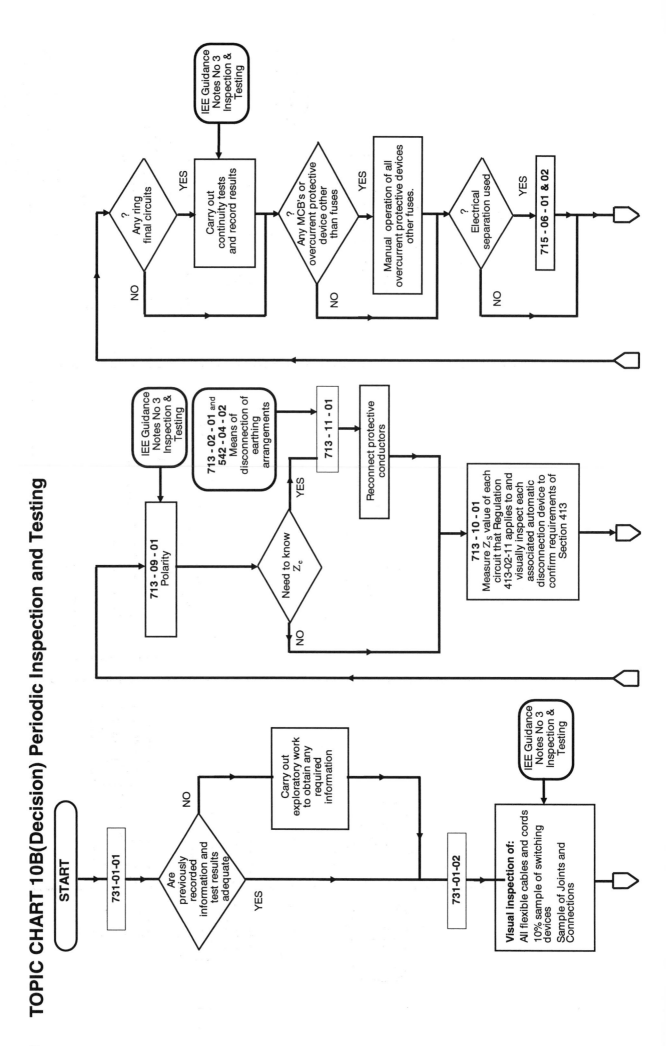

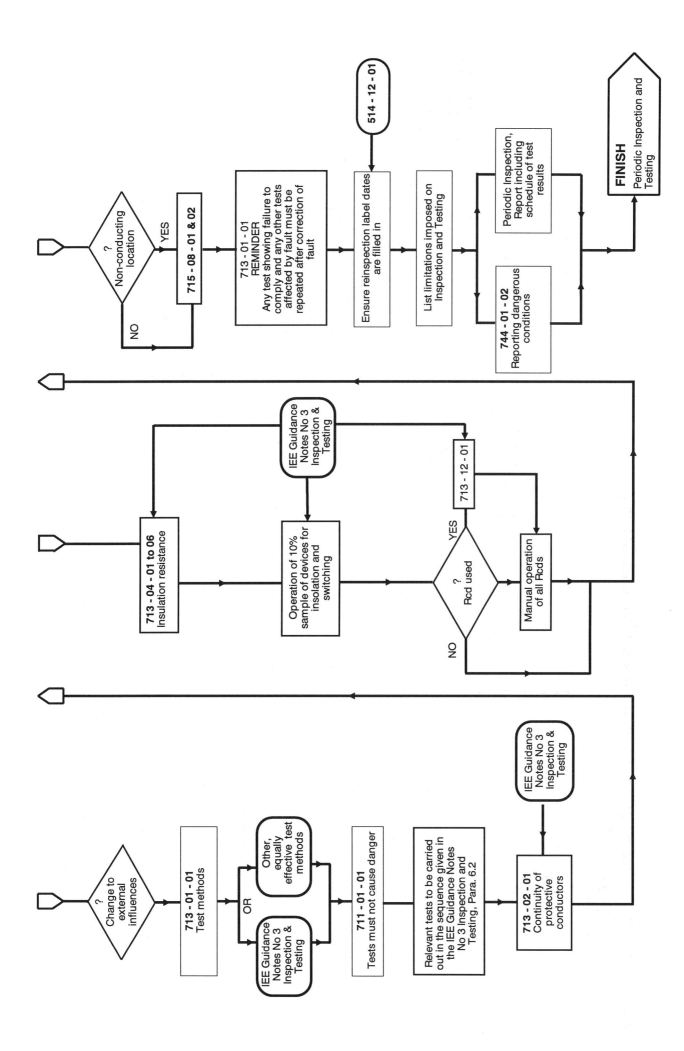

## Documentation and Verification

Before commencing any periodic inspection and testing the available documentation must be gathered together and assessed for compliance with the requirements of Regulation 711-01-02, also to allow testing to be carried out without danger (Regulation 711-01-01).

The Completion Certificate for the initial verification of the original installation and subsequent 'alterations and additions' together with the latest periodic inspection and schedule of test results report should be available for study and to enable a comparison of test results to be made which could indicate an incipient failure of part of the installation.

The prospective fault current at the origin of the installation may have changed due to network modifications by the Regional Electricity Company (REC). The present value should be ascertained by calculation from a measured $Z_e$ or by measurement and compared with the recorded value. If significantly different a report on the required action should be given to the client before proceeding with the Periodic Inspection and Testing.

Any increase in load will have caused an increase in the voltage drop, which should be measured to ensure that there is still compliance with Regulations 525-01-01 and -02.

## Survey

Where the documentation available does not meet the requirements given above an assessment of the shortfall must be made. The Client's agreement should be obtained to the degree of exploratory survey work, disruption to supply and cost to provide the necessary documentation to meet the requirements of future periodic inspection and test. It should be borne in mind that this work could provide much of the information required for the periodic and inspection test being undertaken.

## Report

The format for the periodic inspection and test report is set out in BS 7671;1992 Appendix 6. Forms for reporting and guidance on their completion based on this are available from the ECA, ECA of Scotland and the NICEIC.

For intervals between Periodic Inspections and Tests for Typical Installations (Regulation 732-01-01), see chart on page 145.

The Completion Certificate requires the interval between its issue and the first periodic inspection and testing to be inserted by the person carrying out the test for initial verification. Thereafter each Periodic Inspection report should include the recommended interval to the next Periodic Inspection. These intervals are discretionary based on experience and knowledge of the installation. However, they should not exceed the intervals shown in the following table some of which are mandatory due to statutory instruments or local authority conditions of license.

Those responsible for recommending the intervals must take into account many factors. For example, installations in kitchens of hotels and restaurants should receive careful attention, consideration being given to six-monthly tests. The changing requirements of menus, and the business generally, may lead to the addition of new equipment, which in the absence of the necessary final circuits to supply fixed equipment, may be connected to socket-outlets, with consequent overloading.

Similarly, the introduction of special office equipment may cause overloading of circuits originally intended for relatively small equipment such as typewriters, dictaphones, etc.

Much will depend on the skill and ability of those responsible for the installations, but it is recommended that contractors take care that the duties under the appropriate legislation (Electricity at Work Regulations, 1989, and Health and Safety at Work Act 1974) are brought to the attention of "the duty holder".

Under Regulations 341-01-01 and 732-01-02, it is a requirement that an assessment shall be made, so that any periodic inspection and testing can be carried out safely.

# Intervals Between Periodic Inspections and Tests

| Type of Installation | Reference | Period | M or R* |
|---|---|---|---|
| Domestic | | 10 years | R |
| Commercial | | 5 years or less | R |
| Agricultural and Horticultural | | 3 years | R |
| Temporary installations including construction sites | | 3 months | R |
| Caravan Parks | DOE Model Standards (1977) | 1 year (3 years if underground distribution) | R |
| Fire Alarms | BS5839 Part 1 | 1 year | R |
| Emergency Lighting | BS5266 Part 1 | Detailed recommendations in Section 12 | R |
| Cinema, etc. | Cinematograph (Safety) Regs. 1955 SI 1955 No. 1129 | 1 year | M |
| Petrol Filling Stations | By-laws under Petroleum (Regulation) Acts | 1 year HS (G) 41 recommends annual | M |
| Launderettes | Local Authority By laws | Usually 1 year | M |
| Churches | "Lighting and Wiring of Churches" 1988 | 2 years (new installations) 1 year (installations over 5 years old) | R |
| Industrial | IEE Guidance Notes Inspection and Testing | 3 years | R |
| Leisure Complexes including swimming pools | Local Authority | 1 year | R |
| Places of public entertainment | Local Authority | 1 year if G.S. 50 applies | M |

*M = Mandatory

R = Recommended

Shorter intervals are recommended for tests of operation and of batteries, generators, etc.

N.B. Items noted as "Recommended" may become "Mandatory" when incorporated in Local Authority conditions of licence for premises of various types or may be regarded as such due to the Electricity Supply Regulations 1988 (as amended) or the Electricity at Work Regulations 1989.

# TOPIC CHART 10C (INFORMATION) TESTS AND TEST INSTRUMENTS

## SUPPLY OFF

| REGULATION NUMBER | TESTER | | |
|---|---|---|---|
| | DESCRIPTION | NAME | CHARACTERISTICS |
| 713-02<br><br>713-03 | Continuity of Protective and Final Ring Circuit Conductors | Continuity Tester | O/C volts, **4** volts to 24 volts **D.C. or A.C.** Test current at least **200mA 0.01ohm** resolution 2% accuracy |
| 713-04 & Table 71A | Insulation Resistance | Insulation Tester | Selectable D.C. output voltage of 250V, 500V, 1000V providing 1mA at the selected test voltage, 2% accuracy |
| 713-05 | Site Applied Insulation | Applied Voltage Tester | Variable voltage 0-4000V A.C. Max. output 5mA 5% accuracy |
| 713-06 | Protection by Separation of Circuits | Insulation Tester and Applied Voltage Tester | D.C. output voltage of 500V providing 1mA at the test voltage. 2% accuracy. Applied test voltage 4000V A.C. 5% accuracy |
| 713-07 | Protection by a Site Applied Barrier or Enclosure | BS finger and Test Filament Lamp | Test voltage 40V to 50V |
| 713-08 | Insulation of Non-Conductive Floors and Walls | Insulation Tester | Test voltage 500V D.C. Providing 1mA at the test voltage. 2% accuracy |
| | | Milliammeter | Range 0 to 1mA. 2% max. accuracy with over current protection |
| | | Applied Voltage Tester | Tester voltage 2000V a.c. 5% accuracy |
| 713-09 | Polarity | ContinuityTester; Battery with Voltmeter or Bell Set | O/C volts **4 volts to 24 volts** |
| 713-11 | Earth electrode Resistance | Earth Resistance Tester | 3 or 4 terminal Type. 2% accuracy |

## SUPPLY ON

| REGULATION NUMBER | DESCRIPTION | NAME | CHARACTERISTICS |
|---|---|---|---|
| 713-09 | Polarity | Voltmeter or Filament Lamp or a magnetic indicator | Class 2 BS 89 |
| 713-10 | Earth fault Loop Impedance | Loop Impedance Tester | To provide approx. 25A into the test circuit, 40ms automatic cut off. 0.3 ohm to 25 ohms with 0.01 ohm resolution for up to 50A circuits |
| 713-12 | Operation of RCDs | RCD Tester | Selectable current outputs typically 10mA, 30mA, 100 mA with an accuracy 10% Tripping time indication 50 ms cut off of 5 times operating current. |

**Test Instruments** (see also IEE Guidance Notes No 3 Inspection and Testing)

The undernoted instruments are necessary for testing to BS 7671:1992 requirements:

(1) **Continuity/resistance tester** – this should be a dual range instrument, say 0-1 and 0-10 ohms. Should measurements much below 0.05 ohm be contemplated, then consideration should be given to an instrument which passes a considerable current through the test circuit in order to verify its true value under load or fault conditions.

(2) **Earth electrode resistance tester** - to meet the requirements of BS 7430.

(3) **Insulation resistance tester** – no specific range is laid down, but a centre scale mark of 1 megohm, and a top reading (before infinity) of 50 Megohms would be generally acceptable. Test voltages of 250V, 500V and 1000V will be required.

(4) **High voltage tester** – maximum output 6kV is needed to satisfy 713-05-02(ii).

(5) **Polarity tester** – for checking before the supply is connected, a continuity tester, a battery and voltmeter or a bell-set; for live testing, a magnetic indicator, a voltmeter or a filament lamp (not a neon) with approved fused leads and test prods. N.B. Filament lamps require about 25% voltage to indicate.

(6) **Phase-earth fault impedance testers** - should have a range of at least 0.3 ohm to 25 ohms, (Regulation 413 - 02 - 10 and 11).

(7) **Rcd tester**–capable of indicating that an rcd has operated at its rated residual current, in the time specified in BS 4293; additionally, the RCD should not operate at 50% rated residual current. For a 30mA rcd providing supplementary protection against direct contact, a test current of 150mA should cause the device to operate in not more than 40ms (BS 4293).

## Instruments Generally

It is recommended that instrument identification numbers be recorded against test values to avoid extensive and unnecessary re-testing when one of a number of instruments is found to be faulty. It will also help to locate the particular instrument used, where the test readings are subsequently suspect. Where an instrument has no serial number it should be marked with a unique identification.

A company Quality Assured to BS 5750 must ensure (and for others it is strongly recommended) that all test instruments are subjected to regular checking and recalibration as necessary, and that a proper system of recording and progressing these actions is instituted. The regular use of "check boxes" or test resistors for insulation and continuity instruments is also a useful interim check.

The tests recommended for establishing the continuity of ring circuits are particularly demanding on the instruments used because of the probable low values of resistance involved, and the accuracy required. Hence, the same instrument should be used for the component parts of the testing procedure outlined on page 148, and great care should be taken when "shorting" test leads and adjusting the zero set control on the instrument before taking readings.

It should be remembered that meter accuracy is best in the upper half of the scale; the range giving greatest deflection of the pointer should therefore be selected. Another point to note is that no guidance is given on testing installations in hazardous atmospheres, where the instruments used for testing may themselves be a source of ignition because of the energy stored in the circuits under test. Instruments classified as intrinsically safe are only safe in relation to internal sparking within the test instrument. Precautions must be taken to ensure that the application of the test does not create a hazard.

An expectation of an infinity reading is somewhat meaningless, because a reading of infinity on one instrument design may be shown as a finite reading on an alternative instrument with a different scale or imposed voltage. For a number of mineral insulated cables tested together by an instrument of the type commonly used for insulation testing, a requirement for an infinity reading is over-optimistic. For insulation resistance measurements a value of "infinity" should never be quoted. The reading should always be expressed as being greater than the highest finite scale reading. For example, if the highest scale point is 100 Megohms, but the pointer indicates a greater value, the value should be recorded as "greater than 100 Megohms".

## 1. Continuity of Protective Conductors including Main and Supplementary Equipotential Bonding

### Test Method 1

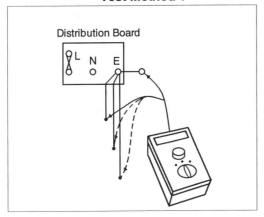

### Test Method 2

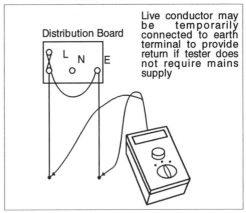

Live conductor may be temporarily connected to earth terminal to provide return if tester does not require mains supply

Tests may be made using either a.c. or d.c. voltages. Regulation 713 - 02 - 01 recommends tests should be made at a voltage more than 4v and no greater than 24v and a short-circuit current not less than 200mA. Where the protective conductor is not steel conduit or a steel enclosure, the requirement for the test current does not apply, and a d.c. ohmmeter may be used, e.g. when the protective conductor is copper.

Test Method (2) should only be used if the protective conductor has negligible inductance.

## 2. Continuity of Ring Final Circuit Conductors

The test method shown below ensures that all conductors of a ring final circuit are continuous and that no interconnecting multiple loops exist

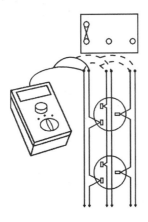

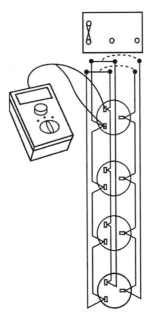

1.  Initial test for continuity. With all ring conductors disconnected from the distribution board measure the resistance of the phase conductor, the neutral conductor and the cpc separately. A finite measurement indicates that the ring conductors are not open circuit.

2.  Compare the measured values obtained in 1. The phase, neutral and cpc should all have the same reading $\pm$ 0.05 ohm if the cpc has the same csa as the phase and neutral conductors.

    If the cpc is a reduced size (1.5mm$^2$) its resistance should be approximately 1.66 times the phase conductor resistance.

3.  Connect the outgoing phase conductor to the return neutral conductor. Connect the return phase conductor to the outgoing neutral conductor.

4.  Measure the resistance between the phase and neutral conductors at each socket outlet.

5.  Repeat 3 and 4 with the phase conductor and the cpc connected.

6.  All measurements at socket outlets on the ring should be approximately equal from each method of connection.

7.  Socket outlets wired as a spur will have proportionately higher values dependent on the length of the spur cable.

8.  The highest recorded resistance will equal $R_1 + R_2$ of the circuit and can be used to determine the earth loop impedance ($Z_S$) to verify compliance with BS 7671:1992 Tables 41B1 and 41B2.

9.  If any significant difference is observed it will indicate a loose connection in a socket outlet or possible wrong identification of single core conductors.

False readings of continuity of ring final circuits which may occur are shown below.

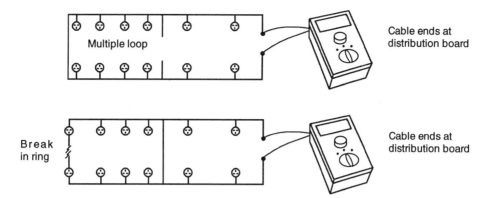

Continuity tests on ring final circuits connected as in (a) and (b) left do not prove continuity of loops, other than that nearest to the distribution board or point of test.

**3. Insulation Resistance** between all Conductors connected to One Phase or Pole, and in turn, all Conductors connected to each Other Phase or Pole.

Note: Any voltage sensitive equipment (e.g. control and instrumentatation equipment incorporating semi-conductor devices) must be disconnected from circuits undergoing an insulation resistance test. Failure to observe this precaution may result in serious damage necessitating costly replacements!

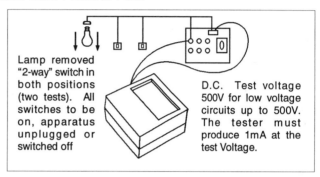

Lamp removed "2-way" switch in both positions (two tests). All switches to be on, apparatus unplugged or switched off

D.C. Test voltage 500V for low voltage circuits up to 500V. The tester must produce 1mA at the test Voltage.

Part 7 of the BS 7671:1992 requires an insulation resistance test to be applied between poles (phase to phase and phase to neutral), and between all poles and earth, (phases and neutral to earth), using a 500V tester for installations rated up to 500V (e.g. 230/400V) and a 1000V tester for installations rated above 500V and up to 1000V.

For these tests, large installations may be divided in accordance with Regulation 713 - 04 - 02. The Regulations require the insulation resistance to be not less than 0.5 Megohm. However for a new installation a value of less than 2 Megohm warrants investigation.

Where possible, current-using equipment should be disconnected and tested separately. The insulation resistance between conductors and exposed conductive parts should be not less than the value given in the relevant British Standard, or where no BS exists, not less than 0.5 megohm.

With filament lamps there is no difficulty with lamp removal, followed by insulation testing between conductors and to earth. Fluorescent and other discharge lighting, may incorporate capacitor discharge resistors across the supply, so that tube removal does not enable insulation tests to be made and in addition electronic components, the presence of which may not be obvious, can be damaged by the test voltage.

It is recommended therefore that such luminaires be individually disconnected before testing final circuits.

## 4. Site Applied Insulation

These tests are only applicable where insulation for protection against direct contact, or supplementary insulation for protection against indirect contact (Class II equipment or equivalent insulation) is applied on site. The tests are not required where factory built and tested equipment is assembled on site.

The tests required by Regulations 713 - 05 - 01 and 02 are of a special nature, and require equipment similar to that used by manufacturers, but not usually available to those engaged in the inspection and testing of electrical installations.

For example, the applied voltage test required by Regulation 713 - 05 - 01 for insulation intended to afford protection against direct contact is required to be equivalent to that specified in the British Standard for similar factory built equipment. For switchgear and control gear, BS 5486 Part 1 is applicable, and where the insulation is rated at between 300V and 660V a.c., the applied voltage test is 2500V a.c. 50 Hz for equipment operating at 200V to 450V.

These tests should be regarded as being of a special nature and not normally included in the programme of testing an installation.

## 5. Protection by Separation of Circuits

Electrical separation is required by Regulation 411 - 02 - 05 (i) where protection against both direct and indirect contact is afforded by SELV and may also be adopted as a method of protection against indirect contact in accordance with Regulations 413 - 06 - 01 to 05.

In both the above examples, it is necessary to verify the degree of electrical separation:

(i)     at the source of supply to the separated circuit,

(ii)    within equipment having live parts connected to normal and separated circuits,

(iii)   between conductors of normal and separated circuits.

Where the supply is derived from a safety isolating transformer complying with BS 3535, it is not necessary to verify item (i) above. In all other instances it must be verified that the source provides an equivalent degree of safety, if necessary by applying the insulation and electric strength tests specified in BS 3535 i.e. 3750 volts a.c. if the working voltage is 200 V to 450 V.

Item (ii) above must also be verified, if necessary by applying the same tests, while item (iii) should be verified as complying with Regulation 411 - 02 - 06 and 413 - 06 - 01(ii) and (iii) as appropriate.

The tests specified in BS 3535 are insulation resistance and electrical strength. The values depend on the type of the Transformer.

## 6. Protection against Direct Contact, by Barriers or Enclosures

Where protection against direct contact is applied on the site by the use of barriers or enclosures, tests must be carried out to verify that these measures meet the requirements of Regulations 413 - 03 - 01 to - 04. These tests are generally a visual examination combined with measurement. For full site verification of IP2X, IPXXB or IP4X special measuring gear such as a B.S. finger may be required, see the IEE Guidance Notes No. 3 Inspection and Testing and BS 5490.

Bearing in mind that these are unusual circumstances, that the use of barriers or enclosures for this purpose is generally in conjunction with a measure for protection against indirect contact, and that the tests require degrees of protection to BS 5490 (see page 124), they are not tests which will normally form part of the testing requirements of an installation.

## 7. Insulation of Non-Conducting Floors and Walls

Whilst Regulation 713 - 08 - 01 specifies that the tests shall be made at not less than three points on the surface, and specifies the relationship of those points to extraneous conductive parts, no indication of the test method is given. Much will depend on the area of the test electrode and the pressure applied to it, as well as the cleanliness of the surfaces in contact.

The nature of the non-conducting situation is such and the details of the test procedures so important, that these tests must be regarded as of a special character requiring the advice of those experienced in such matters, and who will probably be ultimately involved in the use and application of the locations. The distinction should be recognised between non-conducting floors which are intended to insulate, and high resistance floors which are intended to discharge safely any static charges in contact with them.

## 8. Polarity

Polarity testing, although commonly done when the installation is energised, may better be carried out at an earlier stage using a battery and voltmeter or a bell-set. It is not only the connection of switches to the phase conductor which is important, but also fuses and socket-outlets, as well as the centre contacts of Edison screw and single contact bayonet lampholders (where the supply neutral is earthed). Following this, visual inspection is required, noting in particular that all phase conductors are correctly identified (BS 7671:1992 Table 51A).

In general, test equipment relying on neon lamps is unsuitable because of the possibility of erroneous indications, particularly on a.c. circuits. This type of tester should only be used as a secondary or supplementary device; screwdrivers containing neon lamps are not suitable as test instruments.

## 9. Earth Fault Loop Impedance

Where use has to be made of protective measures which require a knowledge of the value of earth-loop impedance, Regulation 713 - 09 - 01 states that the relevant impedances shall be measured, or determined by a suitable alternative method.

During design the maximum earth fault loop impedance ($Z_s$) for the circuits will have been established according to Regulation 413 - 02 - 10 (fuse) or - 11(mcb) (0.4 second disconnection time) or Regulation 413 - 02 - 14 (5 second disconnection time).

If $Z_s$ has been established using the $R_1$ and $R_2$ values from table B1 in the IEE Guidance Notes No 5 - Protection against Electric Shock then $Z_s$ will be related to a conductor temperature of 20°C.

The value of $Z_s$ must meet the requirement of the Regulations at the normal operating temperature of the circuit. The $Z_s$ values obtained from on-site tests do not relate to this condition as they will probably have been taken at ambient temperature. Therefore to be certain of compliance the on-site measurement must be adjusted for the conductor resistance change due to <u>operating</u> conditions and ambient temperature as follows

The earth loop impedance $Z_s$ of a circuit comprises two parts:-

$Z_e$ – external to the circuit, measured from the origin of the circuit via the earth connection

$(R_1 + R_2)$ – the loop impedance of the phase conductor and the cpc before connection to earth, and at 20°C.

During operation, the temperature of the conductor will rise due to the current flow – which will increase the resistance. A factor M is used to correct for this effect. The value of M is dependent on the maximum temperature that the insulation is permitted to attain during operation and must be related to the type of cable being installed or tested. This correction must be used to increase the value of $(R_1 + R_2)$ from the Table B1 value, and is always a multiplier greater than 1. Factor M can be obtained from the specific cable manufacturer or from the formula on page 152.

Thus for calculation for design at 20°C:

$$Z_s = Z_e + M(R_1 + R_2).$$

When $Z_s$ has to be related to ambient temperatures other than 20°C a correction factor C has to be applied to the resistance of the phase conductor and the resistance of the cpc. This factor is a multiplier of less than 1 for temperatures below 20°C and greater than 1 for temperatures above 20°C, thus decreasing the resistance below 20°C or increasing the resistance above 20°C.

Thus for a calculation to define a value for a design incorporating an ambient temperature that is not 20°C:

$$Z_S = Z_e + [M(R_1 + R_2)] \times C_a$$

It should be appreciated however, that the application of this correction factor only amounts to ± 2% for ± 5°C

Measurement of earth fault loop impedance may be carried out using a phase-earth loop tester as described in the IEE Guidance Notes No 3 Inspection and Testing. Neutral-earth loop testers are not recognised for this test, neither are those testers which only give a pass/fail indication because it is necessary to record actual test values.

Before performing an earth loop test, it should be checked that the earthing conductor correctly connects the installation to the means of earthing to avoid the rise in potential on exposed conductive parts which would otherwise occur during the test. The correctness of polarity of the supply must also be verified, and both those test facilities are normally provided in the instrument.

Alternatively, it must be ensured that no person (or livestock) can come into contact with exposed conductive parts for the duration of the testing, e.g. by evacuating the premises.

It should be noted that measurement of $Z_e$ is to be made with the installation disconnected from the means of earthing. Clearly, the main switch of the installation must be in the off position during this test.

To verify compliance with the impedance values given in Tables 41B1, 41B2 and 41D, it is necessary to make two measurements, one at the origin of the installation to determine the external earth fault loop impedance $Z_e$ and the other at the far end of each circuit to determine $Z_s$ the total earth fault loop impedance at ambient temperature. For cables up to 35mm$^2$ csa, (where the inductance of the conductors is negligible compared to their resistance)

$$Z_{s\,(measured)} = Z_e + (R_1 + R_2) \qquad\qquad (1)$$

where $(R_1 + R_2)$ is related to the temperature at which the test is carried out and does not allow for the increase in resistance due to the temperature during operation.

Using the value of $(R_1 + R_2)$ from (1)

$$Z_{S\,(design)} = Z_e + [M(R_1 + R_2)\frac{1}{C_a}] \qquad\qquad (2)$$

The value of $\dfrac{1}{C_a}$ for various ambient temperatures can be obtained from the IEE Guidance Notes No 3 - Inspection and Testing. Typical values for M for cable types are:

| Insulation type | 54B Factor | 54C Factor | |
|---|---|---|---|
| 70°C PVC | 1.04 | 1.2 | 54B applies when CPC is not part of or bunched with cables. |
| 85° Rubber | 1.04 | 1.26 | 54C applies when CPC is incorporated in or bunched with cables |
| 90°C Thermosetting | 1.04 | 1.28 | |

For practical reasons the measurement of $Z_s$ is often made with the main equipotential bonding conductor in place, thus giving a lower reading than would have been obtained if the bond had been removed.

Thus it is often preferable to measure $R_1$ and $R_2$ separately

where $R_1$ is the resistance of the phase conductor (up to 35mm$^2$ csa)
$R_2$ is the resistance of the circuit protective conductor.

The measured values of $R_1, R_2$ and $Z_e$ can then be used in the above formula to calculate $Z_{s\,(design)}$ to compare with tables 41B1, 41B2 and 41D.

The above formula and test are equally applicable for a separate cpc or where the cpc is wholly or mainly conduit or trunking.

To measure $Z_s$ at the most distant lighting point of an installation it is essential to make the test at that ceiling rose. To plug the instrument into a convenient socket outlet and to connect the wander lead to the protective conductor at the lighting point will give low and erroneous readings.

Where the design is to Regulation 413 - 02 - 12 it is always necessary to measure the impedance of the cpc ($R_2$). This value must be corrected for the ambient temperature, checked against the appropriate value in Table 41C and recorded.

$$R_2 \times \frac{1}{C_a}$$

If the phase conductor is larger than 35mm$^2$ $R_1$ is not applicable and a loop impedance test must be carried out.

Where it is necessary to apply an earth fault loop impedance test to an installation or circuit protected by an rcd, it is likely that the test will trip the rcd, especially if its operating current is 30mA or less. To avoid the risks which would result from short-circuiting the rcd, it is recommended that the value should be ascertained by the use of an instrument designed for this purpose or by separate measurements of external and internal impedances.

## 10. Earth Electrode Resistance

The method of testing required under Regulation 713 - 11 - 01 is fully described in BS 7430: 1991 (which supersedes CP 1013 – Earthing).

It is important to appreciate, however, that in this method, the electrode Z(P2) must be in a straight line, between the X(E) and Y(C2) electrodes. In built-up areas, and in certain types of ground, it may be difficult to achieve this and find suitable places to drive the electrodes.

As described in BS 7430 an alternative method is possible by making a measurement of the test electrode resistance in series with another electrode of known negligible resistance, such as a cable sheath or well-earthed water pipe. This method should only be used if the site limitations make it impossible to use the method described in BS 7430, because there will always be a doubt as to whether the resistance areas of the two electrodes are sufficiently independent and, therefore, whether an accurate result is being obtained.

## 11. Functional Testing

Regulation 713 - 12 - 02 requires all components of the installation to be operated to show they are properly mounted, adjusted and installed in accordance with the relevant requirements of BS 7671: 1992.

### Operation of Residual Current Operated Devices

The following are general notes on residual current devices.

**Passive Types**. This type of rcd does not involve amplification of the residual operating current, and consequently does not require the use of electronic circuitry.

Passive types may be separated into two categories: (a) Polarised, and (b) Saturation.

As regards the Polarised category, these are field weakening devices using an armature which is retained in position by a permanent magnet. The excitation current, which is the difference between the flow and return conductor currents at any instant cancels this retaining flux by producing an opposing flux around the transformer.

When the retaining flux is cancelled, the armature is released by springs and the switch mechanism is operated.

Because the flux is affected by an alternating field, it is apparent that it is only weakened during one half of the a.c. cycle. During the other half of the a.c. cycle, the produced flux reinforces the retaining flux of the permanent magnet.

Saturation devices are those which respond to a change in flux pattern of a permanent magnet and core, due to the contributed flux resulting from an imbalance of the windings on the transformer, creating saturation in constricted areas of the laminations. This change of flux causes a weakening of the flux pattern around the armature of the transformer, which is then released by spring operation causing the mechanism to operate. Because saturation can occur during either half of the a.c. cycle, tripping will also occur during either half of the cycle.

**Amplified Type**. This type depends for its successful operation on the correct functioning of an amplifier, the output of which is used to operate a reed relay or other type of relay unit. As it is dependent on a supply voltage for the amplifier to operate, the device will not function if the neutral is lost and will only function within prescribed voltage tolerances. Such devices may also be damaged by the application of high d.c. voltage during testing, and the appropriate precautions should be taken.

Any residual current device provided for protection against indirect contact must be tested by means independent of the test button on the device so as to simulate an appropriate fault condition.

For a residual current device (rcd) the test requirements are given in the IEE Guidance Notes No 3 Inspection and Testing and BS 4293.

Several proprietary rcd testers are available which cause the required residual current to flow from the phase conductor on the load side of the rcd to the associated protective conductor. The test is satisfied if the rated residual current causes the rcd to open within 200 ms, or at any delay time stated by the manufacturer of the rcd.

The tester must not only trip the rcd but must also indicate the time elapsed between the start of the test and the opening of the contacts.

With a fault current flowing equivalent to 50% of the rcd rated tripping current the device should not open. With a 100% rcd rated current flowing the rcd must open in less than 200 ms. If the rcd incorporates an intentional time delay with 100% of the rated time range plus 200 ms.

Due to the variability of time delays a maximum test time cannot be specified (except in the case of the 150 mA test for supplementary protection against direct contact) however it is suggested that a 2 second maximum test time should be sufficient.

If the rcd has been installed as additional protection against direct contact its residual operating current must not exceed 30mA and the test instrument should also be capable of indicating that the rcd has opened within 40ms at a residual current of 150mA, (Regulation 412 - 06 - 02).

After applying the above tests the rcd should be tripped by means of its integral test button to check its effectiveness.

# D

## Explanatory Illustrations

## Explanatory Illustrations

In the Regulations there is much that is not easily described in the form of commentary.

The purpose of the illustrations in this part of the Handbook is to present those requirements visually with full explanatory notes on each.

The illustrations (Figures) contained in the following pages are: Page



Pre
hence for 2.5mm² cable
Imax = 30A
& MCB = 32A. (or fuse)

**The following abbreviations are used in Figure D1 opposite**

R.C. Ring Circuit
S.O. Socket-outlet (BS1363)
S.O.S. As above (with switch)
F.C.U. Fused Connection Unit (max. 13A fuse)
F.C.U.S. As above (with switch)
F.S. Fused Spur
N.F.S. Non-Fused Spur
J.B. Joint Box
S.F.U. Switch & Fuse Unit (13A max.)
C.B. Circuit Breaker (16A max.)

**Figure D1**

# Recommended Arrangement of Domestic Ring Circuits Using Socket-outlets to BS 1363

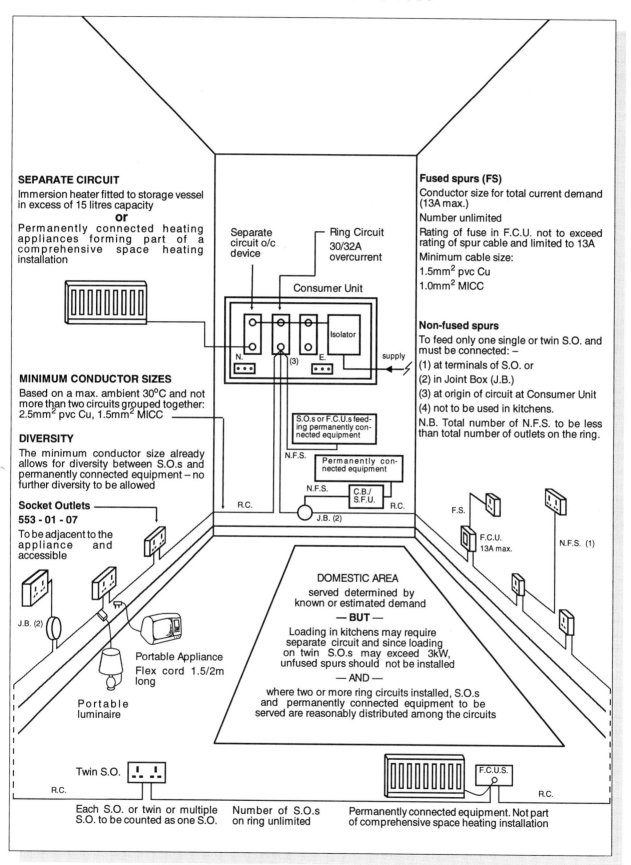

**SEPARATE CIRCUIT**

Immersion heater fitted to storage vessel in excess of 15 litres capacity

**or**

Permanently connected heating appliances forming part of a comprehensive space heating installation

Separate circuit o/c device

Ring Circuit 30/32A overcurrent

Consumer Unit

N.   (3)   E.   supply

Isolator

**MINIMUM CONDUCTOR SIZES**

Based on a max. ambient 30°C and not more than two circuits grouped together: 2.5mm² pvc Cu, 1.5mm² MICC

**DIVERSITY**

The minimum conductor size already allows for diversity between S.O.s and permanently connected equipment – no further diversity to be allowed

**Socket Outlets**
**553 - 01 - 07**

To be adjacent to the appliance and accessible

S.O.s or F.C.U.s feeding permanently connected equipment

N.F.S.

Permanently connected equipment

N.F.S.   C.B./S.F.U.

R.C.   R.C.

J.B. (2)

J.B. (2)

Portable Appliance
Flex cord 1.5/2m long

Portable luminaire

**Fused spurs (FS)**

Conductor size for total current demand (13A max.)

Number unlimited

Rating of fuse in F.C.U. not to exceed rating of spur cable and limited to 13A

Minimum cable size:
1.5mm² pvc Cu
1.0mm² MICC

**Non-fused spurs**

To feed only one single or twin S.O. and must be connected: –

(1) at terminals of S.O. or

(2) in Joint Box (J.B.)

(3) at origin of circuit at Consumer Unit

(4) not to be used in kitchens.

N.B. Total number of N.F.S. to be less than total number of outlets on the ring.

F.S.

F.C.U. 13A max.

N.F.S. (1)

**DOMESTIC AREA**
served determined by known or estimated demand
— **BUT** —
Loading in kitchens may require separate circuit and since loading on twin S.O.s may exceed 3kW, unfused spurs should not be installed
— **AND** —
where two or more ring circuits installed, S.O.s and permanently connected equipment to be served are reasonably distributed among the circuits

Twin S.O.

R.C.

F.C.U.S.

R.C.

Each S.O. or twin or multiple S.O. to be counted as one S.O.

Number of S.O.s on ring unlimited

Permanently connected equipment. Not part of comprehensive space heating installation

# Recommended Arrangements of Radial Circuits, 16A Socket-outlets to BS 4343 1PH & 3PH, 110/240/415V – 50/60Hz

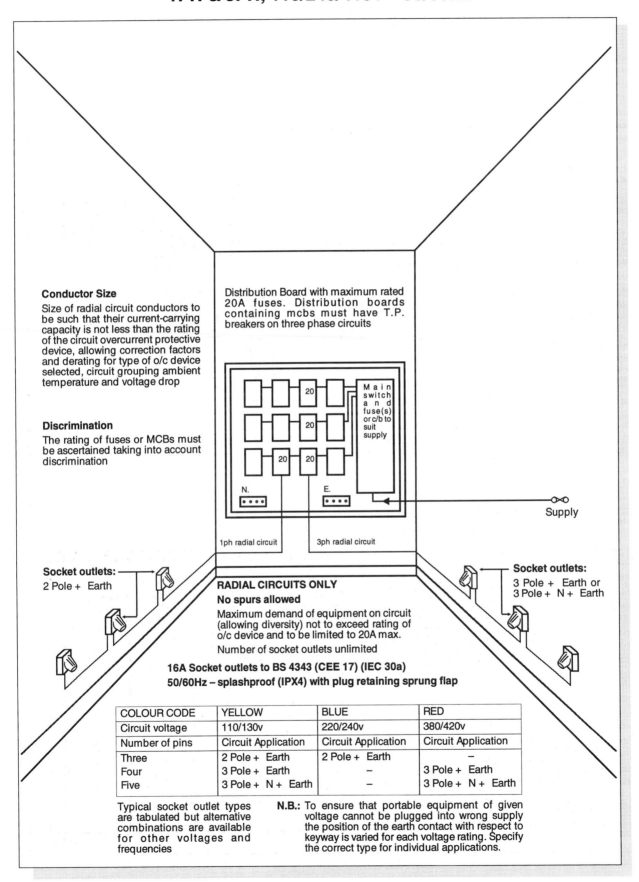

**Conductor Size**

Size of radial circuit conductors to be such that their current-carrying capacity is not less than the rating of the circuit overcurrent protective device, allowing correction factors and derating for type of o/c device selected, circuit grouping ambient temperature and voltage drop

**Discrimination**

The rating of fuses or MCBs must be ascertained taking into account discrimination

Distribution Board with maximum rated 20A fuses. Distribution boards containing mcbs must have T.P. breakers on three phase circuits

20
20
20 20

N.
E.

Main switch and fuse(s) or c/b to suit supply

Supply

1ph radial circuit          3ph radial circuit

**Socket outlets:**
2 Pole + Earth

**Socket outlets:**
3 Pole + Earth or
3 Pole + N + Earth

**RADIAL CIRCUITS ONLY**

**No spurs allowed**

Maximum demand of equipment on circuit (allowing diversity) not to exceed rating of o/c device and to be limited to 20A max.

Number of socket outlets unlimited

**16A Socket outlets to BS 4343 (CEE 17) (IEC 30a)**

**50/60Hz – splashproof (IPX4) with plug retaining sprung flap**

| COLOUR CODE | YELLOW | BLUE | RED |
|---|---|---|---|
| Circuit voltage | 110/130v | 220/240v | 380/420v |
| Number of pins | Circuit Application | Circuit Application | Circuit Application |
| Three | 2 Pole + Earth | 2 Pole + Earth | – |
| Four | 3 Pole + Earth | – | 3 Pole + Earth |
| Five | 3 Pole + N + Earth | – | 3 Pole + N + Earth |

Typical socket outlet types are tabulated but alternative combinations are available for other voltages and frequencies

**N.B.:** To ensure that portable equipment of given voltage cannot be plugged into wrong supply the position of the earth contact with respect to keyway is varied for each voltage rating. Specify the correct type for individual applications.

**Figure D3**

# Recommendations for Cooker Circuits

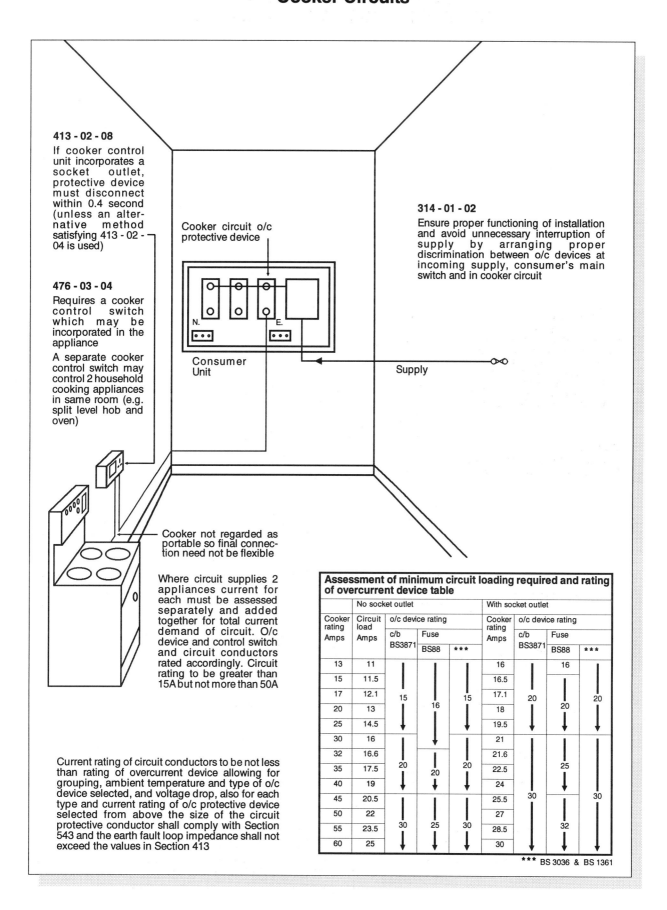

**413 - 02 - 08**

If cooker control unit incorporates a socket outlet, protective device must disconnect within 0.4 second (unless an alternative method satisfying 413 - 02 - 04 is used)

**476 - 03 - 04**

Requires a cooker control switch which may be incorporated in the appliance

A separate cooker control switch may control 2 household cooking appliances in same room (e.g. split level hob and oven)

Cooker circuit o/c protective device

N.     E.

Consumer Unit

**314 - 01 - 02**

Ensure proper functioning of installation and avoid unnecessary interruption of supply by arranging proper discrimination between o/c devices at incoming supply, consumer's main switch and in cooker circuit

Supply

Cooker not regarded as portable so final connection need not be flexible

Where circuit supplies 2 appliances current for each must be assessed separately and added together for total current demand of circuit. O/c device and control switch and circuit conductors rated accordingly. Circuit rating to be greater than 15A but not more than 50A

Current rating of circuit conductors to be not less than rating of overcurrent device allowing for grouping, ambient temperature and type of o/c device selected, and voltage drop, also for each type and current rating of o/c protective device selected from above the size of the circuit protective conductor shall comply with Section 543 and the earth fault loop impedance shall not exceed the values in Section 413

### Assessment of minimum circuit loading required and rating of overcurrent device table

| | No socket outlet | | | | With socket outlet | | | |
|---|---|---|---|---|---|---|---|---|
| Cooker rating Amps | Circuit load Amps | o/c device rating | | | Cooker rating Amps | o/c device rating | | |
| | | c/b BS3871 | Fuse BS88 | *** | | c/b BS3871 | Fuse BS88 | *** |
| 13 | 11 | | | | 16 | | 16 | |
| 15 | 11.5 | | | | 16.5 | | | |
| 17 | 12.1 | 15 | 16 | 15 | 17.1 | 20 | 20 | 20 |
| 20 | 13 | | | | 18 | | | |
| 25 | 14.5 | | | | 19.5 | | | |
| 30 | 16 | | | | 21 | | | |
| 32 | 16.6 | | | | 21.6 | | | |
| 35 | 17.5 | 20 | 20 | 20 | 22.5 | | 25 | |
| 40 | 19 | | | | 24 | | | |
| 45 | 20.5 | | | | 25.5 | 30 | | 30 |
| 50 | 22 | | | | 27 | | | |
| 55 | 23.5 | 30 | 25 | 30 | 28.5 | | 32 | |
| 60 | 25 | | | | 30 | | | |

*** BS 3036 & BS 1361

Page 159

**Figure D4**

# Typical Requirements for Bonding in Locations Containing a Bath Tub or Shower Basin

**Note:**
*In the illustration the only accessible extraneous conductive parts are the towel rail, taps and the bath or wash basin if metal.*

**The illustration shows the towel rail to be water heated, i.e. an extraneous conductive part thus:**

**547 - 03 - 02**

Requires the supplementary bond to be of a conductance not less than half that of the protective conductor connected to the exposed conductive part (the heater) if the bond is mechanically protected, and 4mm² minimum if no mechanical protection is provided.

**If the towel rail is electrically heated it is an exposed conductive part thus:**

**547 - 03 - 01**

Requires the supplementary bond to be of a conductance not less than half that of the smallest protective conductor connected to the exposed conductive part if the bond is mechanically protected, and 4mm² minimum if no mechanical protection is provided

**547 - 03 - 05**

Where a fixed appliance such as a fan heater is supplied by a short length of flexible cord the cpc within the cord can be deemed to satisfy the bonding requirement

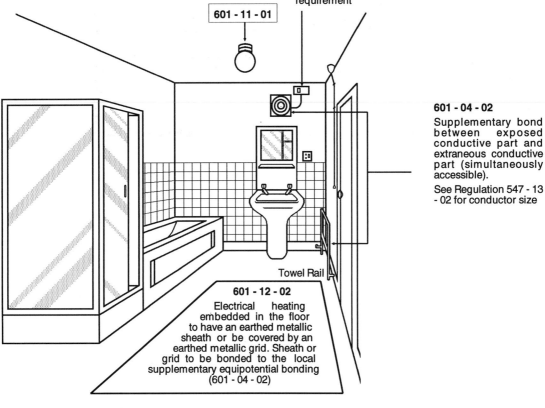

601 - 11 - 01

**601 - 04 - 02**

Supplementary bond between exposed conductive part and extraneous conductive part (simultaneously accessible).

See Regulation 547 - 13 - 02 for conductor size

Towel Rail

**601 - 12 - 02**

Electrical heating embedded in the floor to have an earthed metallic sheath or be covered by an earthed metallic grid. Sheath or grid to be bonded to the local supplementary equipotential bonding (601 - 04 - 02)

**Tests**

These must be carried out to ensure supplementary equipotential bonding is connected to the taps or bath or basin

**601 - 04 - 02**

Irrespective of impedance test results, this Regulation requires supplementary equipotential bonding as follows:-

(1) between simultaneously accessible exposed conductive parts of equipment, e.g. heaters, electric towel rails, electric showers, etc.

(2) between exposed conductive parts and simultaneously accessible extraneous conductive parts

(3) between simultaneously accessible extraneous conductive parts

# E

## Further Information

## Further Information

The various British Standards and Codes of Practice referred to in the Regulations are listed in Appendix 1 of BS 7671:1992. British Standards concerned with Fire Alarms and Emergency Lighting are given in this Section of the Handbook.

Many of the relevant Statutory Regulations and their associated Memoranda are listed in Appendix 2 of BS 7671:1992. This section of the Handbook details more publications which can be of assistance to installation designers and others.

BS 7671: 1992 Amendment 1 incorporates the CENELEC HD 384 series of harmonisation documents listed in the Preface of that BS.

## British Standards Associated with Fire Alarm Systems

BS1635   Graphical Symbols and Abbreviations for Fire Protection Drawings
BS5445   Specifications for Components of Automatic Fire Detection Systems
         Part 1 – Introduction
         Part 5 – Heat Sensitive Detectors
         Part 7 – Point-type Smoke Detectors
         Part 8 - High Temperature Heat Detectors
         Part 9 - Methods of Test of Sensitivity to Fire
BS5446   Specification for Components of Automatic Fire Alarm Systems for Residential Premises
         Part 1 – Point-type Smoke Detectors
BS5839   Fire Detection and Alarm Systems in Buildings
         Part 1 – Code of Practice for System Design, Installation & Servicing
         Part 2 – Specification for Manual Call Points
         Part 3 – Specification for Automatic Release Mechanisms for Certain Fire Protection Equipment
         Part 4 – Specification for Control and Indicating Equipment
         Part 5 – Specification for Optical Beam Smoke Detectors

## British Standards and Codes of Practice Associated with Emergency Lighting

BS764   Automatic Change-over Contactors
BS5266   Emergency Lighting
         Part 1 – Code of Practice for Premises other than Cinemas
         Part 2 – will replace CP1007 – Maintained Lighting for Cinemas
         Part 3 – Small Power Relays
CP1007   Maintained Lighting for Cinemas

## Statutory Regulations, Associated Memoranda and Other Information

Appendix 2 of BS 7671: 1992 lists a number of relevant statutory regulations and their associated memoranda, whilst Appendix 1 lists British Standards and Codes of Practice referred to in the Regulations. There are various other publications which can be of assistance to installation designers and others, these include:

**Association of British Theatre Technicians, 47 Bermondsey Street, London SE1 3XT, Tel: 0171-403 3778 Fax: 0171-378 6170**

Safe Sound - guide to electrical safety for musicians

**The Brewers Society, 42 Portman Square, W1H OBB, 0171-486 4831**

Code of Practice for Electrical Safety in Beer Dispense in Licenced Premises (1990)

**British Approvals Service for Electrical Equipment in Flammable Atmospheres. (BASEEFA)**

BASEEFA LIST 1993: Certified and approved electrical equipment. 1989
BASEEFA technical rules new series (TRNS) No.1-14

A complete list of BASEEFA publications is available from BASEEFA, Health and Safety Executive, Harpur Hill, Buxton, Derbyshire SK17 9JN


**Building Services Research and Information Association, Old Bracknell Lane West, Bracknell, Berks RG12 4AH, 01344 426511**

Application guide 1/81 – Locating Fire Alarm Sounders for Audibility
Technical Note 6/76 – Approximation of the Electrical Load of Lighting Installations


**The Chartered Institution of Building Services Engineers, Delta House, 222 Balham High Road, London, SW12 9BS, 0181-675 5211**

CIBSE Guides
Code for Interior Lighting
Lighting in Hostile and Hazardous Conditions
LG1:        The Industrial Environment
LG2:        Hospitals and Health Care Buildings
LG3:        Areas for Visual Display Terminals
LG4:        Sports
LG5:        Lecture, Teaching and Conference Rooms
LG6:        The Outdoor Environment
LG7:        Lights for Offices
LG8:        Museums and Art Galleries
TM12:       Emergency Lighting
TM16:       Fire Precautions – Sources of Information on Legal and other Services


Prefix LG is for a lighting guide. A full list of all CIBSE publications is obtainable from the CIBSE bookshop at the above address.

**Church House Bookshop, Great Smith Street, Westminster, London SW1T 9NZ, 0171-222 5520**

Lighting and Wiring of Churches

**Copper Development Association, Potters Bar, Herts. 01707 650711**

Copper for Busbars
Copper for Earthing
Copper for Electrical Purposes

**The Electrical Contractors' Association Security Group, ESCA House, 34 Palace Court, Bayswater, London W2 4HY, 0171-229 1266** or
**The Electrical Contractors' Association of Scotland Security Group, Bush House, Bush Estate, Midlothian, EH26 0SB, 0131-445 5577**

Code of Practice for Security Installations
Code of Practice for Controlled Access Installations
Code of Practice for CCTV Installations
Code of Practice for Security Installations in all Type of Vehicles/Boats excluding ships

**The Electricity Association, 30 Millbank, London SW1P 4RD, 0171-834 2333**

The Electricity Association publish a great many well illustrated publications, which are always being added to. They include:

Information on lighting in all aspects and environments
Guidance on heat pumps, air conditioning etc.
Data sheets and case histories on a wide variety of industrial processes
Booklets on electric space and water heating for industrial, commercial and industrial applications, process heating, furnaces, etc., materials handling and electric vehicles

Engineering Recommendations

| | |
|---|---|
| G5/3 (1976) | Limits for harmonics in the U.K. electricity systems |
| G12/2 (1982) | National Code of Practice on the application of protective multiple earthing on low voltage networks |
| G15 (1971) | Standardisation of electricity supply arrangements in industrialised component buildings |
| G16 (1972) | Standard specification for a new single phase service termination/metering up to 100A |
| G22/2 (1984) | Superimposed signals on public electricity supply networks |
| G59 (1985) | Recommendations for the connection of private generating plant to the electricity boards' distribution systems |
| P9 (1963) | Supply to welding plant and 1984 Amendment regarding d.c. welders |
| P13/1 (1979) | Electric motors - starting conditions |
| P22 (1981) | Procedure for advising customers on the unsuitability of water pipes for use as earth electrodes |
| P23 (1981) | Consumers' earth fault protection of single phase supplies up to 100A for compliance with the 15th Edition of the IEE Wiring Regulations for electrical installations. |
| P26 (1985) | The short circuit characteristics of electricity boards' low voltage distribution networks and the co-ordination of overcurrent protective devices of 240V supplies up to 100A |
| P25 (1984) | The estimation of the maximum prospective short circuit for three-phase 415V supplies |

ACE Reports

| | |
|---|---|
| ACE 7 (1963) | Supply to welding plant |
| ACE 15 (1970) | Harmonic distortion caused by converter equipment |
| ACE 22 (1969) | Survey on the amplitude and frequency of occurrence of surges on M.V. systems |
| ACE 41 (1975) | Guide to an ERA Code of Practice for the avoidance of interference with electronic instrumentation and systems |
| ACE 73 (1969) | Limits for harmonics in the U.K. electricity supply systems |

**Employment Department, Moorfoot, Sheffield, S14 PQ 0114 275 3275**

Safety Recommendations for on-the-job Training in Electrical Skills

**Fire Protection Association, 140 Aldersgate Street, EC1A 4HX, 0171-606 3757**

Automatic Fire Alarm Installations for Protection of Property. 1985
Information on static electricity fire risks
Various booklets on electrical fire hazards

**Health and Safety Executive, Rose Court, 2 Southwark Bridge, London, SE1 9HS, 0171 717 600**

Publications can be obtained from HSE Books, PO Box 1999, Sudbury, Suffolk CO10 6FS, 01787 881165.

As mentioned in the introduction, HSE have a wide range of explanatory and advisory booklets and pamphlets in addition to the statutory publications. Those listed below are in the order in which they appear in the HSE "Publications in Series List", and are of particular relevance to electrical contractors, but the full list should be consulted for further information. The full list is published yearly and a subscription service for information on new or amended publications is available from the HSE.

| | |
|---|---|
| AS17 | Electricity on the Farm |
| CS2 | Storage of highly flammable liquids |
| GS6(Rev.) | Avoidance of danger from overhead electrical lines |
| GS23 | Electrical safety in schools |
| GS24 | Electricity on construction sites |
| GS27 | Protection against electric shock |
| GS34 | Electrical safety in departments of electrical engineering |
| GS37 | Flexible leads, plugs, sockets, etc. |
| GS38(Rev.) | Electrical test equipment for use by electricians |
| HS(G)13 | Electrical testing: safety in electrical testing |
| HS(G)22 | Electrical apparatus for use in potentially explosive atmospheres |
| HS(G)38 | Lighting at work |
| HS(G)41 | Petrol Filling Stations: Construction and operation |
| HS(G)47 | Avoiding danger from underground services |
| HS(R)25 | Memorandum of Guidance on Electricity at Work Regulations |
| IND(G)56(P) | Flammable Liquids on Construction Sites |
| PM32(Rev.) | Safe use of portable electrical apparatus (electrical safety) |
| PM37 | Electrical installations in motor vehicle repair premises and amendment sheet |
| PM38 | Selection and use of electric handlamps |
| PM53 | Emergency private generation: electrical safety |
| PM64 | Electrical safety in arc welding |
| SS2 | Safe Use of Ladders |
| SS3 | General Access Scaffolds |
| SS6 | Use of portable electric tools and equipment on construction sites |
| SS7 | Avoiding danger from buried services |
| SS8 | Safety in excavations. |
| SS10 | Tower scaffolds |
| SS11 | Safe Use of Propane and LPG cylinders |

Occasional Papers

| | |
|---|---|
| OP10 | Safety of Electrical Distribution Systems on Factory Premises |

Unnumbered Reports

| | |
|---|---|
| – | Programmable electronic systems in safety related applications: an introductory guide. 1987 |
| – | Programmable electronic systems in safety related applications: general technical guidelines. 1987 |

**Her Majesty's Stationery Office, 49 High Holborn, London WC1, 0171-873 9090**

Electric Lighting (Clauses) Act, 1899
Electricity Supply Acts, 1882 to 1928
Electricity Act, 1947
Electricity Supply (Meters) Act, 1936, and Permitted Alterations Act 1958
Highly Flammable Liquids & Liquified Petroleum Gas Regulations 1972
Gas Safety Regulations 1988
List of Certificates for Intrinsically Safe and Approved Electrical Apparatus for use in certain
    specified atmospheres
The Electricity Supply Regulations 1988 as amended
The Electricity at Work Regulations 1989
Managing Health and Safety in Construction: Principles and Application to Main Contractor/
    Sub-Contractor Projects. 1987 (HMSO 0118839896)

**Home Office Fire Department, Headquarters, Queen Anne's Gate, SW1H 9AT, 0171-273 2427**

Fire Safety Management in Hotels and Boarding Houses
Draft Guide to Fire Precautions in Hospitals
Guide to Fire Precautions in Existing Places of Entertainment and Like Premises
Guide to Fire Precautions in Existing Hotels & Boarding Houses that Require a Fire Certificate

**Institution of Electrical Engineers, Savoy Place, London WC2R 0BL, 0171-240 1871**

IEE Guidance Notes
1.  Selection & Erection
2.  Isolation & Switching
3.  Inspection & Testing
4.  Protection Against Fire
5.  Protection Against Electric Shock
6.  Protection Against Overcurrent
On-Site Guide

**The Lighting Industry Federation Ltd., Swan House, 207 Balham High Road, London SW17 7BQ, 0181-675 5432**

Interior Lighting Design Handbook 6th Edition
Energy Managers Lighting Handbook
High Frequency Ballasts for Tubular Fluorescent Lamps
Lighting Controls and Energy Management Systems: Applications Guide
(PCBs) Lighting Guide
Better Lighting at Work

| LIF3 | Factfinder | Lamp Guide |
| LIF4 | Factfinder | Lighting and Energy |
| LIF5 | Factfinder | Benefits of Certification [ICEL] |
| LIF6 | Factfinder | Hazardous Area Lighting |

**The Loss Prevention Council, 140 Aldersgate Street, London EC1A 4HY, 0171-606 1050**

LPC Rules for Automatic Sprinkler Installations
Plus a range of other publications

**National Engineering Specification Ltd., Southgate Chambers, 37/39 Southgate St., Winchester SO23 9EH, 01962-842058**

The National Engineering Specification

**National Inspection Council for Electrical Installation Contracting, Vintage House, 36/37 Albert Embankment, London SE1 7UJ, 0171-582 7746**

Snags and Solutions

**NHS Estates, Department of Health**, 1 Trevelyan Square, Boar Lane, Leeds LS1 6AE, 0113 254 700

Hospital Technical Memoranda
No.      1 – Anti-static precautions: rubber, plastics and fabrics
No.      2 – Anti-static precautions, flooring in anaesthetising areas
No.     82 – Fire Alarm And Detection Sytems
No. 2007 – Electrical Services: supply and distribution
No. 2011 – Emergency Electrical Services
No. 2014 – Abatement of Electrical Interference
No. 2015 – Patient/nurse call systems
No. 2017 – Engineering Installations: Commissioning and associated activities
No. 2020 – Electrical Safety Standard – LV Systems
No. 2021 – Electrical Safety Standard – HV Systems

**Office of Telecommunications - OFTEL, Export House, 50 Ludgate Hill, EC4M 7JJ 2HQ, 071-634 8764**

Telecommunication Wiring in Business Premises and Houses

**Royal Society for the Prevention of Accidents, Occupational Safety Section, Head Office, Cannon House, Priory Queensway, Birmingham B4 6BS, 0121-200 2461**

The Safe Use of Electricity
Range of safety publications, notices and training aids

**Telecommuncations Users' Association, 48 Percy Road, London, N12 8BU 0181-445 0996**

Telecom Users' Handbook

# F

## Index to Handbook

## Index to Handbook

References are to page numbers in the Handbook